IAENG TRANSACTIONS ON ENGINEERING SCIENCES

Special Issue for the International Association of
Engineers Conferences 2019

IAENG TRANSACTIONS ON ENGINEERING SCIENCES

Special Issue for the International Association of Engineers Conferences 2019

Editors

Sio-Iong Ao
International Association of Engineers, Hong Kong

Haeng Kon Kim
Daegu Catholic University, South Korea

Oscar Castillo
Tijuana Institute of Technology, Mexico

Alan Hoi-shou Chan
City University of Hong Kong, Hong Kong

Hideki Katagiri
Kanagawa University, Japan

NEW JERSEY • LONDON • SINGAPORE • BEIJING • SHANGHAI • HONG KONG • TAIPEI • CHENNAI • TOKYO

Published by

World Scientific Publishing Co. Pte. Ltd.
5 Toh Tuck Link, Singapore 596224
USA office: 27 Warren Street, Suite 401-402, Hackensack, NJ 07601
UK office: 57 Shelton Street, Covent Garden, London WC2H 9HE

British Library Cataloguing-in-Publication Data
A catalogue record for this book is available from the British Library.

IAENG TRANSACTIONS ON ENGINEERING SCIENCES
Special Issue for the International Association of Engineers Conferences 2019

ISBN 978-981-121-508-7

For any available supplementary material, please visit
https://www.worldscientific.com/worldscibooks/10.1142/11687#t=suppl

PREFACE

A large international conference on Advances in Engineering Sciences was held in Hong Kong, 13-15 March, 2019, under the International MultiConference of Engineers and Computer Scientists 2019 (IMECS 2019). The IMECS 2019 is organized by the International Association of Engineers (IAENG). IAENG is a non-profit international association for the engineers and the computer scientists, which was founded originally in 1968 and has been undergoing rapid expansions in recent few years. The IMECS conference serves as a good platform for the engineering community to meet with each other and to exchange ideas. The conference has also struck a balance between theoretical and application development. The conference committees have been formed with over three hundred committee members who are mainly research center heads, faculty deans, department heads, professors, and research scientists from over 30 countries with the full committee list available at our conference web site (http://www.iaeng.org/IMECS2019/committee.html). The conference is truly an international meeting with a high level of participation from many countries. The response that we have received for the conference is excellent. There have been more than four hundred manuscript submissions for the IMECS 2019. All submitted papers have gone through the peer review process and the overall acceptance rate is 50.36%.

This volume contains twelve revised and extended research articles written by prominent researchers participating in the conference. Topics covered include engineering physics, computer science, electrical engineering, industrial engineering, and industrial applications. The book offers the state of art of tremendous advances in engineering sciences and engineering technologies and applications, and also serves as an excellent reference work for researchers and graduate students working with/on engineering sciences and engineering technologies.

Sio-Iong Ao
Haeng Kon Kim
Oscar Castillo
Alan Hoi-shou Chan
Hideki Katagiri

CONTENTS

Preface v

Mobile Manipulation for Human-Robot Collaboration in Intralogistics 1
Christof Röhrig and Daniel Heß

A Simple Baseline Correction Method for Raman Spectra 21
Yung-Sheng Chen and Yu-Ching Hsu

Biquadratic Allpass Phase-Compensating System Design Utilizing Bilinear Error 33
Tian-Bo Deng

Road Usage Recovery in Fukushima Prefecture Following the 2011 Tohoku Earthquake 48
Jieling Wu and Noriaki Endo

Sentiment Analysis on Twitter with Python's Natural Language Toolkit and VADER Sentiment Analyzer 63
Shihab Elbagir and Jing Yang

Analyzing the Tendency of Review Expressions for the Automatic Construction of Cosmetic Evaluation Dictionaries 81
Mayumi Ueda, Yuna Taniguchi, Yuki Matsunami, Asami Okuda and Shinsuke Nakajima

A Web Search Method Based on Complaint Data Analysis for Collecting Advice 96
Liang Yang, Daisuke Kitayama and Kazutoshi Sumiya

LOD-Based Semantic Web for Indonesian Cultural Objects 108

Gloria Virginia, Budi Susanto and Umi Proboyekti

Age Difference of Self Learning Aspect, Effective Learning Aspect and Multimedia Aspect Towards E-Learning Courses in Hong Kong Higher Education 128

H.K. Yau and K.C. Yeung

Design the Process Separation of Methylal/Methanol by Response Surface Methodology Based on Process Simulation in Extractive Distillation 140

W. Weerachaipichasgul, A. Wanwongka, S. Saengdaw and A. Chanpirak

Update to Strong and Weak Force Proposed Equations 154

Matthew Cepkauskas

Effect of Bandgap Vs Electron Tunneling on Photovoltaic Performance of Ringwood (Syzgium Anisatum) Dye-sensitized Solar cell 167

T.J. Abodunrin, O.O. Ajayi, M.E. Emetere, A.P.I. Popoola, U.O. Uyor and O. Popoola

Mobile Manipulation for Human-Robot Collaboration in Intralogistics

Christof Röhrig* and Daniel Heß+

University of Applied Sciences and Arts in Dortmund,
Institute for the Digital Transformation of Application and Living Domains,
Otto-Hahn-Str. 23, 44227 Dortmund, Germany,
Web: www.fh-dortmund.de/en/idial/, *Email:* *roehrig@ieee.org,
+daniel.hess@fh-dortmund.de

In human-robot collaboration, robots interact directly and safely with humans in a shared workspace without protective guards. This paper presents the design of *OmniMan*, which is a mobile manipulator for human-robot collaboration consisting of an omnidirectional mobile platform and a lightweight robotic arm. Mobile manipulators extend the workspace of manipulators (robotic arms) by mounting them on mobile platforms. The mobile platform is driven by Mecanum wheels, which provide 3 degrees of freedom in motion. The paper develops the kinematic model of the whole system including platform and manipulator. Furthermore the overall controller structure for *OmniMan* including the real-time synchronization of platform and manipulator is described.

Keywords: Mecanum; Mobile Manipulator; Human Robot Collaboration

1. Introduction

In human-robot collaboration humans interact with robots or manipulators in a shared workspace without protective guards. Manipulators have been used in a large range of applications mainly in the production industry, but their application is limited in scenarios which need a very large working space such as aerospace manufacturing, ship building, and wind turbine manufacturing. Mobile manipulation is a solution to overcome these limitations and counts to be a key technology not only for the production industry but also for professional service robotics such as intralogistics. Furthermore mobile manipulators find their applications in virtual or real laboratories for research and education[1,2].

In the last years, research institutes have developed their own mobile manipulators based on commercial available mobile platforms and robotic arms. *Cody* from Georgia Institute of Technology consists of two arms from MEKA Robotics and a Mecanum wheeled Segway platform[3]. Cody was build mainly for research on service robotics in the health care domain. Another example in this area is *POLAR* (PersOnaL Assistant Robot) from Cornell University, which consists of a 7 DoF (degrees of freedom) Barrett arm mounted on a Segway Omni base[4]. *TOMM* from TU Munich is a wheeled humanoid robot, which consists of a Mecanum wheeled mobile Platform and two UR5 robotic arms[5]. Other popular examples in this domain are Willow Garage's PR2 and the Fraunhofer IPA's Care-O-bot 4.

Compared to the domains of professional and domestic service, mobile manipulators for industrial applications operate in more structured environments. However, they require a higher level of operational efficiency, e.g. in terms of speed, accuracy and robustness, to be suitable for industrial applications[6]. Popular examples of mobile manipulators for industrial environments are KUKA's *omniRob* and *Moiros*[7] as well as *ANNIE* and *LiSA* from Fraunhofer IFF, which are all based on Mecanum wheeled platforms. *omiRob* is used mainly in intralogistics applications[8–10], where human-robot collaboration is often needed.

A Mecanum wheel consists of a central hub with free moving rollers, which are usually mounted at $\pm 45°$ angles around the hubs' periphery. The outline of the rollers is such that the projection of the wheel appears to be circular. A mobile platform driven by Mecanum wheels provides 3 degrees of freedom in motion. Usually such platforms consists of four or more wheels. A typical configuration is the four-wheeled one of the *URANUS* omnidirectional mobile robot[11]. The drive structure of a Mecanum wheeled mobile platform with four or more wheels is over-actuated, which means that actuation conflicts may occur.

This paper extends the work presented in[12]. The kinematic equations of a mobile manipulator including 6- DoF manipulator and omnidirectional platform are developed. The paper develops a generalized kinematic model of an over-actuated Mecanum wheeled mobile platform, which includes the kinematic motion constraints of the system and that is valid for Omni wheels as well. Furthermore the paper describes the overall controller structure of our mobile manipulator *OmniMan* (Fig. 1). *OmniMan* consists of an Universal Robots UR5 robotic arm mounted on Mecanum wheeled mobile platform from MIAG Fahrzeugbau GmbH (Braunschweig, Germany). The paper describes different options for interfacing an UR arm controller to a central controller PC as well as to synchronize the movements of the arm with the platform in real-time.

Figure 1. Mobile manipulator *OmniMan*: UR5 robotic arm on a Mecanum wheeled mobile platform

2. Problem Formulation

We consider the problem of motion control of a mobile manipulator, which consists of a robotic arm and a mobile platform. The robotic arm provides 6 DoF in 3D Cartesian space, the omnidirectional mobile platform provides additional 3 DoF in 2D space (position and heading). Therefore, the omnidirectional mobile manipulator is an over-determined system, where the additional degrees of freedom in configuration space can be used to optimize the movement of the manipulator. In general, the forward kinematics of a mobile manipulator can be described by

$$\boldsymbol{x} = \boldsymbol{f}(\boldsymbol{q}), \text{ with } \boldsymbol{x} = [x, y, z, \alpha, \beta, \gamma]^{\mathrm{T}}, \text{ and } \boldsymbol{q} = \begin{pmatrix} \boldsymbol{q}_{\mathrm{p}} \\ \boldsymbol{q}_{\mathrm{a}} \end{pmatrix}, \tag{1}$$

where $\boldsymbol{x}$ is the pose of the mobile manipulator's tool frame with respect to the world frame ($\boldsymbol{x} \in \mathbb{R}^3 \times \mathrm{SO}(3)$), $\boldsymbol{q}$ are the generalized coordinates in the configuration space (C-space) with the 2D pose of the mobile platform $\boldsymbol{q}_{\mathrm{p}} =$

$[x_p, y_p, \theta_p]^T$, $\boldsymbol{x}_p \in \mathbb{R}^2 \times \mathbb{S}^1$ and the joint angles of the arm $\boldsymbol{q}_a = [\theta_1, \ldots, \theta_6]^T$. The inverse kinematics of the system

$$\boldsymbol{q} = \boldsymbol{f}^{-1}(\boldsymbol{x}) \tag{2}$$

is over-determined and can be solved by numerical methods in order to optimize the motion in C-space.

The mobile platform is driven by 4 Mecanum wheels. A Mecanum wheeled mobile platform provides any desired motion in $x-$ and $y-$direction and rotation θ around the $z-$axis, simultaneously. The mobile platform moves in 2D space, the pose of the platform in the world frame is defined as $\boldsymbol{x}_p = (x_p, y_p, \theta_p)^T$ (see Fig. 2). The robot frame is a coordinate

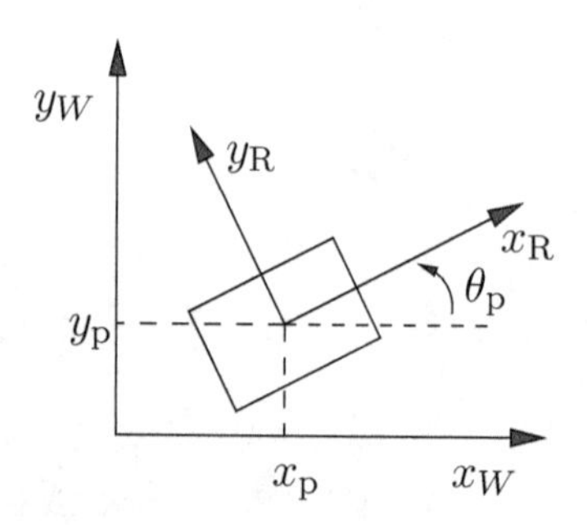

Figure 2. Definition of pose $\boldsymbol{x}_p = (x_p, y_p, \theta_p)^T$ and robot frame

frame, which is fixed at the mobile platform. Velocities in the robot frame (F_R) can be transformed into the world frame (F_W), as a function of the heading of the platform ($\theta = \theta_p$):

$$\dot{\boldsymbol{x}}_R = \boldsymbol{R}(\theta)\,\dot{\boldsymbol{x}}_W, \quad \Rightarrow \quad \dot{\boldsymbol{x}}_W = \boldsymbol{R}^{-1}(\theta)\,\dot{\boldsymbol{x}}_R \tag{3}$$

$$\text{with} \quad \dot{\boldsymbol{x}}_W = \begin{pmatrix} \dot{x}_W \\ \dot{y}_W \\ \dot{\theta} \end{pmatrix}, \dot{\boldsymbol{x}}_R = \begin{pmatrix} \dot{x}_R \\ \dot{y}_R \\ \dot{\theta} \end{pmatrix}, \boldsymbol{R}(\theta) = \begin{pmatrix} \cos\theta & \sin\theta & 0 \\ -\sin\theta & \cos\theta & 0 \\ 0 & 0 & 1 \end{pmatrix}$$

The kinematic model of mobile platforms equipped with n Mecanum wheels is well known (see[11,13]). The inverse kinematics of the platform, which transform the velocity of the platform into wheel velocities, can be described by linear equations in the robot frame. It can be written under the general matrix form:

$$\dot{\boldsymbol{\varphi}} = \boldsymbol{J}_w \dot{\boldsymbol{x}}_R, \text{ with } \boldsymbol{J}_w \in \mathbb{R}^{n\times 3}, \tag{4}$$

where $\dot{\boldsymbol{\varphi}} = (\dot{\varphi}_1, \dot{\varphi}_2, \ldots, \dot{\varphi}_n)^{\mathrm{T}}$ are the angular velocities of the wheels and $\boldsymbol{J}_{\mathrm{w}}$ is a Jacobi matrix with constant parameters. For a platform equipped with $n > 3$ Mecanum wheels, the forward kinematic equations are over-determined. The forward kinematics can be obtained by using a least square approach and applying the Moore-Penrose pseudo-inverse to $\boldsymbol{J}$:

$$\dot{\boldsymbol{x}}_{\mathrm{R}} = \boldsymbol{J}_{\mathrm{w}}^{+} \dot{\boldsymbol{\varphi}}, \text{ with } \boldsymbol{J}_{\mathrm{w}}^{+} = (\boldsymbol{J}_{\mathrm{w}}^{\mathrm{T}} \boldsymbol{J}_{\mathrm{w}})^{-1} \boldsymbol{J}_{\mathrm{w}}^{\mathrm{T}} \tag{5}$$

In contrast to the overdeterminacy in the mobile manipulator, the over-actuated drive structure of the platform can not be used to achieve additional degrees of freedom in C-space. The overdeterminacy of the platform yields to motion constraints, which must be fulfilled to avoid additional wheel slippage.

3. Kinematic Model of the Mobile Manipulator

3.1. *Kinematic Model of the Universal UR5*

The UR5 is a lightweight 6 DoF industrial manipulator manufactured by Universal Robots (UR). It has a weight of 18.4 kg, a reach of 85 cm and a maximal payload of 5 kg [14].

The forward kinematics of the UR5 arm can be easily described by the Denavit-Hartenberg (DH) method. The homogeneous transformation associated with one link can be described by

$$\begin{aligned} {}^{i-1}\boldsymbol{T}_i &= \begin{pmatrix} {}^{i-1}\boldsymbol{R}_i & {}^{i-1}\boldsymbol{p}_i \\ \boldsymbol{0} & 1 \end{pmatrix} \\ &= \begin{pmatrix} \cos\theta_i & -\sin\theta_i \cos\alpha_i & \sin\theta_i \sin\alpha_i & a_i \cos\theta_i \\ \sin\theta_i & \cos\theta_i \cos\alpha_i & -\cos\theta_i \sin\alpha_i & a_i \sin\theta_i \\ 0 & \sin\alpha_i & \cos\alpha_i & d_i \\ 0 & 0 & 0 & 1 \end{pmatrix} \end{aligned} \tag{6}$$

where ${}^{i-1}\boldsymbol{R}_i \in \mathrm{SO}(3)$ represents the orientation, ${}^{i-1}\boldsymbol{p}_i \in \mathbb{R}^3$ the position, θ_i, d_i, a_i, α_i are the DH parameters. The DH parameters of the UR5 arm can be found in Fig. 3. The forward kinematics of a 6 DoF robotic arm can be described by

$${}^{0}\boldsymbol{T}_6 = \prod_{i=1}^{6} {}^{i-1}\boldsymbol{T}_i\,, \tag{7}$$

where ${}^{i-1}\boldsymbol{T}_i$ are the DH matrices. The forward kinematics of the robotic

i	θ_i	d_i	a_i	α_i
1	θ_1	0,08920	0	$+90°$
2	θ_2	0	0,425	0
3	θ_3	0	0,392	0
4	θ_4	0,10930	0	$-90°$
5	θ_5	0,09475	0	$+90°$
6	θ_6	0,08250	0	0

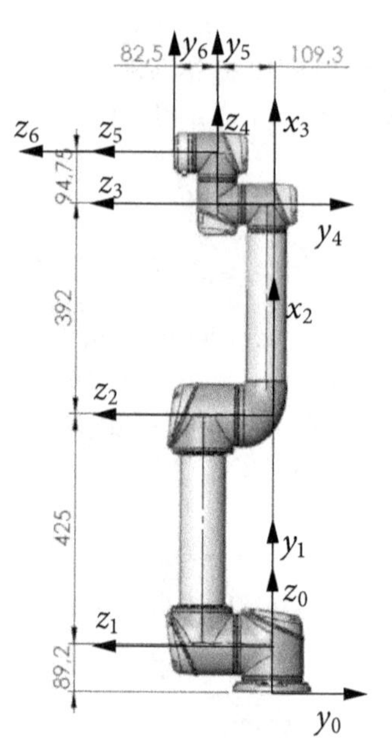

Figure 3. UR5 robotic arm with DH frames and DH parameters

arm can be described in compact form:

$$\boldsymbol{x} = \boldsymbol{f}_{\mathrm{a}}(\boldsymbol{q}_{\mathrm{a}}), \text{ with } \boldsymbol{x} = [x, y, z, \alpha, \beta, \gamma]^{\mathrm{T}}, \boldsymbol{q}_{\mathrm{a}} = [\theta_1, \ldots, \theta_6]^{\mathrm{T}}, \quad (8)$$

where $\boldsymbol{x}$ is the pose of the robotic arm in Cartesian space and $\boldsymbol{q}_{\mathrm{a}}$ are the joint variables of the arm. The inverse kinematics describe the joint angles as a function of the pose of the robotic arm

$$\boldsymbol{q}_{\mathrm{a}} = \boldsymbol{f}_{\mathrm{a}}^{-1}(\boldsymbol{x}), \quad (9)$$

and can not be solved in analytic form in general. For the Universal UR5 arm, an analytic solution for the inverse kinematics was developed by Kelsey Hawkins[15]. We solve the inverse kinematics of the mobile manipulator using a numerical solution, in order to use the additional degrees of freedom of the mobile manipulator for optimization of the movement in C-space. The numerical solution uses the Jacobian

$$\boldsymbol{J}_{\mathrm{a}}(\boldsymbol{q}) = \frac{\partial \boldsymbol{f}_{\mathrm{a}}(\boldsymbol{q}_{\mathrm{a}})}{\partial \boldsymbol{q}_{\mathrm{a}}} = \begin{pmatrix} \frac{\partial f_1}{\partial \theta_1} & \frac{\partial f_1}{\partial \theta_2} & \cdots & \frac{\partial f_1}{\partial \theta_6} \\ \vdots & \vdots & \ddots & \vdots \\ \frac{\partial f_6}{\partial \theta_1} & \frac{\partial f_6}{\partial \theta_2} & \cdots & \frac{\partial f_6}{\partial \theta_6} \end{pmatrix}. \quad (10)$$

The i-th column of $\boldsymbol{J}_\mathrm{a}$ can be obtained numerically

$$\boldsymbol{J}_{\mathrm{a},i} = \begin{pmatrix} \boldsymbol{J}_\mathrm{v} \\ \boldsymbol{J}_\omega \end{pmatrix} = \begin{pmatrix} \boldsymbol{z}_{i-1} \times (\boldsymbol{p}_6 - \boldsymbol{p}_{i-1}) \\ \boldsymbol{z}_{i-1} \end{pmatrix}, \tag{11}$$

where $\boldsymbol{p}_6$ is the position of the TCP. The vectors $\boldsymbol{p}_{i-1}$ and $\boldsymbol{z}_{i-1}$ can be found in the DH matrix

$${}^{0}\boldsymbol{T}_{i-1} = \prod_{j=1}^{j=i-1} {}^{j-1}\boldsymbol{T}_j = \begin{pmatrix} \boldsymbol{x}_{i-1} & \boldsymbol{y}_{i-1} & \boldsymbol{z}_{i-1} & \boldsymbol{p}_{i-1} \\ 0 & 0 & 0 & 1 \end{pmatrix}. \tag{12}$$

3.2. *Motion Model with Constraints for Mobile Platforms with Mecanum Wheels*

In this section, the kinematics of an omnidirectional mobile platform with n Mecanum wheels is developed. A Mecanum wheel consists of a central hub with free moving rollers. A mobile platform driven by 3 or more Mecanum wheels provides 3 DoF in motion. For wheel i, we define the wheel frame F_i and the roller frame $F_{\mathrm{r},i}$, which are in a fixed position in the robot frame F_R (see Fig. 4). For simplification, the kinematic model is described in 2D, which means that all frames lie within the xy-plane. The position of the wheel frame F_i with respect to the robot frame is described by the 3 constant parameters: α_i, l_i and δ_i, where δ_i defines the rotation angle between F_i and the F_R and is usually equal to zero. γ_i defines the angle of the roller with respect to the wheel frame. $\varphi_i(t)$ drives the wheel and defines the rotation angle of the wheel around its horizontal axis of rotation. The wheel is driven in the direction of it's x_i axis. The wheel has one contact point to the plane and is able to rotate around this point (rotation around its z_i axis).

For simplification, it is assumed that there is only one roller that has contact to the floor and that the contact point stays always in the center of the roller (and the wheel). The roller frame $F_{\mathrm{r},i}$ as well as the wheel frame F_i has its origin in this point of contact. The $x_{\mathrm{r},i}$ axis lies in the shaft of the roller. The wheel is able to move free in direction of the $y_{\mathrm{r},i}$ axis.

The movements of the platform yield to the velocities $\dot{x}_\mathrm{R}$, $\dot{y}_\mathrm{R}$, $l \cdot \dot{\theta}$ in the contact point of the wheel, which can be transformed to the roller frame $F_{\mathrm{r},i}$:

$$\dot{x}^\mathrm{R}_{\mathrm{r},i} = \dot{x}_\mathrm{R} \cos(\gamma_i + \delta_i) + \dot{y}_\mathrm{R} \sin(\gamma_i + \delta_i) + l_i \, \dot{\theta} \, \cos(\alpha_i + {}^{\pi}\!/_2 - \delta_i - \gamma_i) \tag{13}$$

$$\dot{y}^\mathrm{R}_{\mathrm{r},i} = -\dot{x}_\mathrm{R} \sin(\gamma_i + \delta_i) + \dot{y}_\mathrm{R} \cos(\gamma_i + \delta_i) + l_i \, \dot{\theta} \, \sin(\alpha_i + {}^{\pi}\!/_2 - \delta_i - \gamma_i) \tag{14}$$

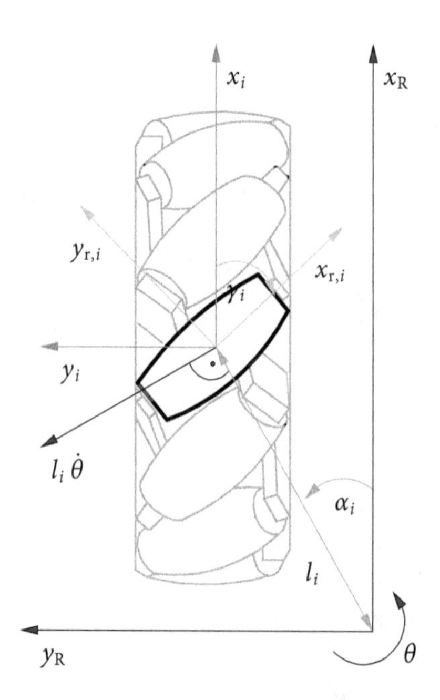

Figure 4. Mecanum wheel with frames

where γ_i is the rotation angle between the wheel frame and the roller frame (usually $\pm 45°$) and δ_i is the angle between the wheel frame and the robot frame (usually $0°$). The rotation of the wheel drives the velocity $\dot{x}_i = r \cdot \dot{\varphi}_i$ in the point of contact, which can be transformed in x and y components of the roller frame $K_{\mathrm{r},i}$:

$$\dot{x}^{\varphi}_{\mathrm{r},i} = r\,\dot{\varphi}_i\,\cos(\gamma_i), \quad \dot{y}^{\varphi}_{\mathrm{r},i} = -r\,\dot{\varphi}_i\,\sin(\gamma_i) \tag{15}$$

Since the roller can not move in direction of its shaft, $\dot{x}^{\mathrm{R}}_{\mathrm{r},i} = \dot{x}^{\varphi}_{\mathrm{r},i}$ must be true, which leads to

$$\dot{x}_{\mathrm{R}}\cos(\gamma_i+\delta_i)+\dot{y}_{\mathrm{R}}\sin(\gamma_i+\delta_i)+l_i\,\dot{\theta}\,\cos(\alpha_i+\pi/2-\delta_i-\gamma_i) = r\,\dot{\varphi}_i\,\cos(\gamma_i), \tag{16}$$

and finally to the inverse kinematic equation of wheel i

$$\dot{\varphi}_i = \frac{1}{r \cdot \cos(\gamma_i)}\big(\cos(\delta_i+\gamma_i),\ \sin(\delta_i+\gamma_i),\ l_i\sin(\delta_i+\gamma_i-\alpha_i)\big)\dot{\boldsymbol{x}}_{\mathrm{R}}. \tag{17}$$

For a platform equipped with n Mecanum wheels, (17) can be used to obtain the inverse kinematics of the platform as

$$\dot{\boldsymbol{\varphi}} = \boldsymbol{J}_{\mathrm{w}}\dot{\boldsymbol{x}}_{\mathrm{R}}, \text{ with } \boldsymbol{J}_{\mathrm{w}} \in \mathbb{R}^{n\times 3} \tag{18}$$

If γ_i is set equal to zero, the Mecanum wheel becomes an Omni wheel and therefore (17) can be used to model the inverse kinematics of an Omni wheel driven platform as well.

For a platform with more than 3 Mecanum wheels, the kinematics are over-determined, which means that the wheel speed of $m = n - 3$ of the n wheels are a linear combination of the other ones. Only 3 rows of J_{w} are linear independent, m rows are a linear combination of the first 3 rows. To obtain the kinematic motion constraints, J_{w} can be split up into 2 sub-matrices:

$$\boldsymbol{J}_{\mathrm{w}} = \begin{pmatrix} \boldsymbol{J}_3 \\ \boldsymbol{J}_m \end{pmatrix}, \text{ with } \boldsymbol{J}_3 \in \mathbb{R}^{3\times 3}, \boldsymbol{J}_m \in \mathbb{R}^{m\times 3}, m = n - 3 \tag{19}$$

The velocities in the robot frame can be defined by 3 wheels:

$$\begin{pmatrix} \dot{x}_{\mathrm{R}} \\ \dot{y}_{\mathrm{R}} \\ \dot{\theta} \end{pmatrix} = \boldsymbol{J}_3^{-1} \begin{pmatrix} \dot{\varphi}_1 \\ \dot{\varphi}_2 \\ \dot{\varphi}_3 \end{pmatrix}$$

this leads to m kinematic motion constraints:

$$\boldsymbol{J}_m \boldsymbol{J}_3^{-1} \begin{pmatrix} \dot{\varphi}_1 \\ \dot{\varphi}_2 \\ \dot{\varphi}_3 \end{pmatrix} = \begin{pmatrix} \dot{\varphi}_4 \\ \vdots \\ \dot{\varphi}_n \end{pmatrix},$$

which can be expressed in general matrix form

$$\boldsymbol{T}\dot{\boldsymbol{\varphi}} = \boldsymbol{0}, \text{ with } \boldsymbol{T} = \left(\boldsymbol{J}_m \boldsymbol{J}_3^{-1}, -\boldsymbol{I}_m\right), \tag{20}$$

where $\boldsymbol{I}$ denotes the identity matrix. We define a vector of angular error velocities $\dot{\boldsymbol{\epsilon}} = (\dot{\epsilon}_1 \ldots \dot{\epsilon}_m)^{\mathrm{T}}$ to detect violations of the kinematic constraints:

$$\dot{\boldsymbol{\epsilon}} = \boldsymbol{T}\dot{\boldsymbol{\varphi}}, \tag{21}$$

$\boldsymbol{T}$ and $\dot{\boldsymbol{\epsilon}}$ can be used to augment the kinematics of the platform

$$\dot{\boldsymbol{x}}_{\mathrm{g}} = \boldsymbol{J}_{\mathrm{g}}^{-1}\dot{\boldsymbol{\varphi}}, \text{ with } \dot{\boldsymbol{x}}_{\mathrm{g}} = \begin{pmatrix} \dot{\boldsymbol{x}}_{\mathrm{R}} \\ \dot{\boldsymbol{\epsilon}} \end{pmatrix}, \boldsymbol{J}_{\mathrm{g}}^{-1} = \begin{pmatrix} \boldsymbol{J}_{\mathrm{w}}^{+} \\ \boldsymbol{T} \end{pmatrix} \tag{22}$$

where $\boldsymbol{J}_{\mathrm{g}}$ is an invertible square matrix, which describes the augmented inverse kinematics of the system:

$$\dot{\boldsymbol{\varphi}} = \boldsymbol{J}_{\mathrm{g}}\dot{\boldsymbol{x}}_{\mathrm{g}} \tag{23}$$

The sub-vector $\dot{\boldsymbol{\epsilon}}$ in $\dot{\boldsymbol{x}}_{\mathrm{g}}$ can be used to control the kinematic constraints.

A typical configuration of a Mecanum wheeled platform consists of 4 wheels (see Fig. 5). The positions of the wheels with respect to the robot

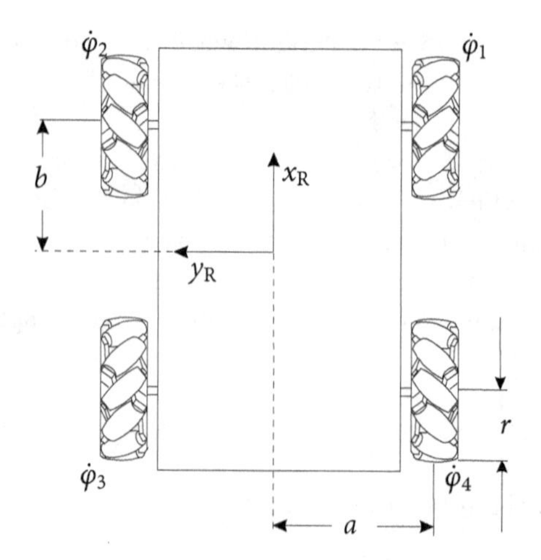

Figure 5. Omnidirectional platform with Mecanum wheels (top view)

Table 1. Configuration of the Platform in Fig. 5

i	α_i	γ_i
1	$-\arctan\left(\frac{a}{b}\right)$	$+\frac{\pi}{4}$
2	$+\arctan\left(\frac{a}{b}\right)$	$-\frac{\pi}{4}$
3	$\frac{\pi}{2}+\arctan\left(\frac{a}{b}\right)$	$+\frac{\pi}{4}$
4	$-\frac{\pi}{2}-\arctan\left(\frac{a}{b}\right)$	$-\frac{\pi}{4}$

frame can be defined as $\delta_i = 0$ and $l_i = \sqrt{a^2+b^2}$. This leads to the configuration parameters shown in Table 1. These parameters in in conjunction with (17) yield to the inverse kinematics

$$\begin{pmatrix}\dot{\varphi}_1\\ \dot{\varphi}_2\\ \dot{\varphi}_3\\ \dot{\varphi}_4\end{pmatrix} = \boldsymbol{J}_{\mathrm{w}} \begin{pmatrix}\dot{x}_{\mathrm{R}}\\ \dot{y}_{\mathrm{R}}\\ \dot{\theta}\end{pmatrix}, \text{ with } \boldsymbol{J}_{\mathrm{w}} = \frac{1}{r}\begin{pmatrix}1 & 1 & (a+b)\\ 1 & -1 & -(a+b)\\ 1 & 1 & -(a+b)\\ 1 & -1 & (a+b)\end{pmatrix} \tag{24}$$

The kinematic constraint can be obtained using (20)

$$0 = \boldsymbol{T}\dot{\boldsymbol{\varphi}}, \text{ with } \boldsymbol{T} = \left(1, 1, -1, -1\right), \tag{25}$$

which leads to the augmented forward kinematics

$$\dot{\boldsymbol{x}}_{\rm g} = \boldsymbol{J}_{\rm g}^{-1}\dot{\boldsymbol{\varphi}}, \text{ with } \dot{\boldsymbol{x}}_{\rm g} = \begin{pmatrix} \dot{x}_{\rm R} \\ \dot{y}_{\rm R} \\ \dot{\theta} \\ \dot{\epsilon}_1 \end{pmatrix}, \dot{\boldsymbol{\varphi}} = \begin{pmatrix} \dot{\varphi}_1 \\ \dot{\varphi}_2 \\ \dot{\varphi}_3 \\ \dot{\varphi}_4 \end{pmatrix},$$
$$\boldsymbol{J}_{\rm g}^{-1} = \frac{r}{4}\begin{pmatrix} 1 & 1 & 1 & 1 \\ 1 & -1 & 1 & -1 \\ \frac{1}{a+b} & \frac{-1}{a+b} & \frac{-1}{a+b} & \frac{1}{a+b} \\ \frac{4}{r} & \frac{4}{r} & -\frac{4}{r} & -\frac{4}{r} \end{pmatrix} \tag{26}$$

and the augmented inverse kinematics

$$\dot{\boldsymbol{\varphi}} = \boldsymbol{J}_{\rm g}\dot{\boldsymbol{x}}_{\rm g}, \text{ with } \boldsymbol{J}_{\rm g} = \frac{1}{r}\begin{pmatrix} 1 & 1 & (a+b) & 1 \\ 1 & -1 & -(a+b) & 1 \\ 1 & 1 & -(a+b) & -1 \\ 1 & -1 & (a+b) & -1 \end{pmatrix} \tag{27}$$

where r is the radius of the wheels, a and b are given by the dimension of the platform (see Fig. 5). Eqn. (26) is used in the motion controller of the platform to execute odometry and to sense the coupling error. Eqn. (27) is used in the motion controller to control the speeds in robot frame and to control the kinematic constraints.

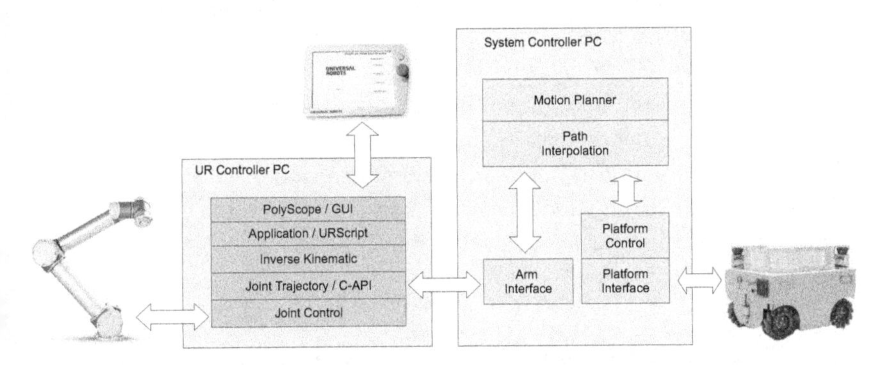

Figure 6. Control architecture of Mecanum wheeled mobile manipulator

3.3. *Experimental Results*

In order to evaluate the effectiveness of the coupling controller several experiments are performed with some of our mobile platforms, which are shown

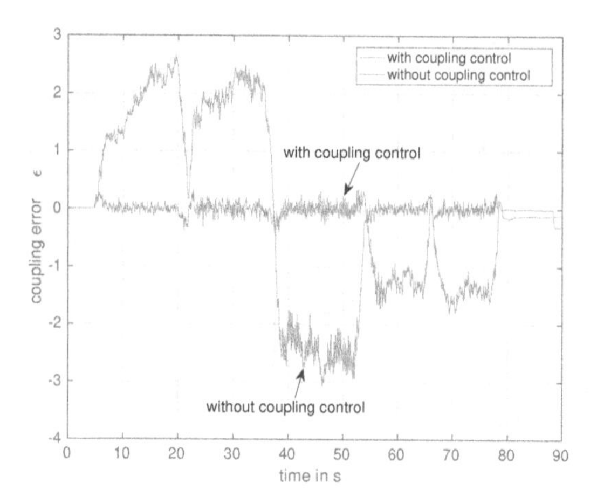

Figure 7. Coupling error over time

in Fig 1. The results of one of these experiments are shown in Fig. 7. The platform moves a path with straight lines forwards, sideways and backwards. After that, it moves a circular path with two rotations back to the starting point. The red curve in the picture shows the coupling error ϵ_1. The five movements of the platform yield to a coupling error different from zero, where the individual movements can be easily distinguished. The coupling error may be caused by parameter differences between the velocity control loops of individual wheels for instance variations in friction, motor constants or amplifier gains. The coupling error leads to an additional wheel slippage, which decreases the accuracy of odometry.

The blue curve in Fig 7 shows the coupling error ϵ_1 with active coupling control ($k_c = 2$). Compared to the red curve, the coupling error is reduced nearly to zero.

3.4. *Dynamic Model for Mobile Platforms with Mecanum Wheels*

This section will briefly describe a dynamic model of Mecanum wheeled platforms including the friction force of the Mecanum wheels. The dynamics of a Mecanum wheeled platform can be described in the robot frame:

$$\boldsymbol{F}_{\mathrm{R}} = \boldsymbol{H}\,\ddot{\boldsymbol{x}}_{\mathrm{R}} = \boldsymbol{J}^{-1}\,\boldsymbol{F}_{\mathrm{wheel}} \tag{28}$$

with

$$\boldsymbol{F}_{\mathrm{R}} = \begin{pmatrix} F_{\mathrm{x}} \\ F_{\mathrm{y}} \\ M_{\theta} \end{pmatrix}, \quad \boldsymbol{H} = \begin{pmatrix} m & 0 & 0 \\ 0 & m & 0 \\ 0 & 0 & I \end{pmatrix}, \quad \boldsymbol{F}_{\mathrm{wheel}} = \begin{pmatrix} F_1 \\ F_2 \\ \vdots \\ F_n \end{pmatrix},$$

where $\boldsymbol{F}_{\mathrm{R}}$ are the translational forces and the moment of force (torque) in robot frame, which accelerate the platform, m is the mass of the platform for translation, I is the moment of inertia for platform rotation around the z axis and $\boldsymbol{F}_{\mathrm{wheel}}$ are the tangential contact forces of the Mecanum wheels, which accelerate the platform. The acceleration in robot frame can be obtained from accelerations in world frame by differentiating (3), which leads to

$$\ddot{\boldsymbol{x}}_{\mathrm{R}} = \dot{\boldsymbol{R}}(\theta)\,\dot{\boldsymbol{x}}_{\mathrm{W}} + \boldsymbol{R}(\theta)\,\ddot{\boldsymbol{x}}_{\mathrm{W}} \tag{29}$$

and considers the effect of the *Coriolis* force. Each wheel i is driven by an electrical motor, which produces the moment

$$M_i = r\,F_i + M_{\mathrm{fric},i} + I_i\,\ddot{\varphi}_i \tag{30}$$

where r is the wheel radius, M_{fric} is the total of the moment produced by friction, I_i is the moment of inertia of the wheel and $\ddot{\varphi}_i$ is the angular acceleration of the wheel: $\ddot{\boldsymbol{\varphi}} = \boldsymbol{J}\ddot{\boldsymbol{x}}_{\mathrm{R}}$ The moment of motor i is produced by the armature current of the motor: $M_i = c_{\phi}\,I_{\mathrm{a},i}$, where c_{ϕ} is the motor constant. The dynamics of the motor current can be described as

$$\begin{aligned} L_{\mathrm{a}}\,\frac{\mathrm{d}I_{\mathrm{a},i}}{\mathrm{d}\,t} + R_{\mathrm{a}}\,I_{\mathrm{a},i} &= U_{\mathrm{a},i} - U_{\mathrm{emf},i}, \\ \text{with} \quad U_{\mathrm{emf},i} &= c_{\phi}\,n_{\mathrm{gear}}\,\dot{\varphi}_i \end{aligned} \tag{31}$$

where L_{a} is the armature inductance, R_{a} is the armature resistance, U_{a} is the armature voltage, U_{emf} is the voltage produced by the back electro mechanical force, n_{gear} is the ratio of the gear that connects motor with wheel.

The moment produced by friction M_{fric} is a sum of the internal friction produced by the motor and the gear, the rolling friction on the floor and the friction in the bearing of the rollers and therefore M_{fric} depends largely on the direction of movement. We model the moment produced by friction as a function of the platform movement

$$M_{\mathrm{fric},i} = f_i(\dot{\boldsymbol{x}}_{\mathrm{R}}), \tag{32}$$

which can be experimentally identified[16].

3.5. *Forward Kinematics of the Mobile Manipulator*

The development of the forward kinematics is straight forward. The position of the platform with respect to the world frame can be described in 3D by the homogeneous transformation matrix

$$ {}^{\mathrm{w}}\boldsymbol{T}_{\mathrm{p}}(\boldsymbol{q}_{\mathrm{p}}) = \begin{pmatrix} \cos\theta_{\mathrm{p}} & -\sin\theta_{\mathrm{p}} & 0 & x_{\mathrm{p}} \\ \sin\theta_{\mathrm{p}} & \cos\theta_{\mathrm{p}} & 0 & y_{\mathrm{p}} \\ 0 & 0 & 1 & 0 \\ 0 & 0 & 0 & 1 \end{pmatrix} \tag{33} $$

where θ_{p} is the heading of the platform with respect to the world frame and $(x_{\mathrm{p}}, y_{\mathrm{p}})$ is the position of the platform in the world frame.

The robotic arm is mounted on the platform in the fixed position $(d_{\mathrm{x}}, d_{\mathrm{y}}, d_{\mathrm{z}})$ with respect to the robot frame, which yields to the constant homogeneous transformation matrix

$$ {}^{\mathrm{P}}\boldsymbol{T}_0 = \begin{pmatrix} 1 & 0 & 0 & d_{\mathrm{x}} \\ 0 & 1 & 0 & d_{\mathrm{y}} \\ 0 & 0 & 1 & d_{\mathrm{z}} \\ 0 & 0 & 0 & 1 \end{pmatrix} . \tag{34} $$

The forward kinematics of the robot arm are described by the DH-matrix

$$ {}^{0}\boldsymbol{T}_6(\boldsymbol{q}_{\mathrm{a}}) = \prod_{i=1}^{6} {}^{i-1}\boldsymbol{T}_i , \tag{35} $$

this yields directly to the forward kinematics $\boldsymbol{x} = \boldsymbol{f}(\boldsymbol{q})$ of the whole mobile manipulator, which can be obtained from

$$ {}^{\mathrm{w}}\boldsymbol{T}_6(\boldsymbol{q}_{\mathrm{p}}, \boldsymbol{q}_{\mathrm{a}}) = {}^{\mathrm{w}}\boldsymbol{T}_{\mathrm{p}}(\boldsymbol{q}_{\mathrm{p}}) \cdot {}^{\mathrm{P}}\boldsymbol{T}_0 \cdot {}^{0}\boldsymbol{T}_6(\boldsymbol{q}_{\mathrm{a}}) . \tag{36} $$

3.6. *Inverse Kinematics of the Mobile Manipulator*

We solve the inverse kinematics of the mobile manipulator

$$ \boldsymbol{q} = \boldsymbol{f}^{-1}(\boldsymbol{x}) $$

using a numerical approach. This approach needs the Jacobian matrix of the forward kinematics of the mobile manipulator:

$$ \boldsymbol{J}(\boldsymbol{q}) = \frac{\partial \boldsymbol{f}(\boldsymbol{q})}{\partial \boldsymbol{q}} = \begin{pmatrix} \frac{\partial f_1}{\partial q_1} & \frac{\partial f_1}{\partial q_2} & \cdots & \frac{\partial f_1}{\partial q_n} \\ \vdots & \vdots & \ddots & \vdots \\ \frac{\partial f_6}{\partial q_1} & \frac{\partial f_6}{\partial q_2} & \cdots & \frac{\partial f_6}{\partial q_n} \end{pmatrix} $$

This Jacobian consists of the Jacobian of the mobile platform and the manipulator

$$\dot{\boldsymbol{x}} = \boldsymbol{J}(\boldsymbol{q}) \cdot \dot{\boldsymbol{q}}, \text{ with } \boldsymbol{J}(\boldsymbol{q}) = (\boldsymbol{J}_{\mathrm{p}}(\boldsymbol{q}), \boldsymbol{J}_{\mathrm{a}}(\boldsymbol{q}))\,. \tag{37}$$

The $\boldsymbol{J}_{\mathrm{a}}(\boldsymbol{q})$ can be obtained from (10), (11), (12). The 3D kinematics of the platform can be obtained from (33) and (34)

$$\boldsymbol{x} = \boldsymbol{f}_{\mathrm{p}}(\boldsymbol{q}_{\mathrm{p}}) = \begin{pmatrix} x_{\mathrm{p}} + d_{\mathrm{x}} \cos\theta_{\mathrm{p}} - d_{\mathrm{y}} \sin\theta_{\mathrm{p}} \\ y_{\mathrm{p}} + d_{\mathrm{x}} \sin\theta_{\mathrm{p}} + d_{\mathrm{y}} \cos\theta_{\mathrm{p}} \\ d_{\mathrm{z}} \\ 0 \\ 0 \\ \theta_{\mathrm{p}} \end{pmatrix},$$

$$\text{with } \boldsymbol{x} = [x, y, z, \alpha, \beta, \gamma]^{\mathrm{T}}\,,\ \boldsymbol{q}_{\mathrm{p}} = [x_{\mathrm{p}}, y_{\mathrm{p}}, \theta_{\mathrm{p}}]^{\mathrm{T}}, \tag{38}$$

where $(d_{\mathrm{x}}, d_{\mathrm{y}}, d_{\mathrm{z}})$ denotes the position of the arm on the platform. Thereby, the Jacobian is easily obtained:

$$\boldsymbol{J}_{\mathrm{p}}(\boldsymbol{q}_{\mathrm{p}}) = \frac{\partial \boldsymbol{f}_{\mathrm{p}}(\boldsymbol{q}_{\mathrm{p}})}{\partial \boldsymbol{q}_{\mathrm{p}}} = \begin{pmatrix} 1 & 0 & -d_{\mathrm{x}} \sin\theta_{\mathrm{p}} - d_{\mathrm{y}} \cos\theta_{\mathrm{p}} \\ 0 & 1 & d_{\mathrm{x}} \cos\theta_{\mathrm{p}} - d_{\mathrm{y}} \sin\theta_{\mathrm{p}} \\ 0 & 0 & 0 \\ 0 & 0 & 0 \\ 0 & 0 & 0 \\ 0 & 0 & 1 \end{pmatrix} \tag{39}$$

In order to solve the inverse kinematics, a rough estimation $\boldsymbol{q}_k$, near the final solution $\boldsymbol{q} = \boldsymbol{f}^{-1}(\boldsymbol{x})$ is needed. If the mobile manipulator moves a continues path, this can be the preceding point on the path. Starting with $\boldsymbol{q}_k$, joint coordinates nearer to the final solution can be obtained by

$$\boldsymbol{q}_{k+1} = \boldsymbol{q}_k + \boldsymbol{J}^{-1}(\boldsymbol{q}_k) \cdot (\boldsymbol{x} - \boldsymbol{x}_k), \text{ with } \boldsymbol{x}_k = \boldsymbol{f}(\boldsymbol{q}_k)$$

The iteration is aborted, when the error falls below a predefined limit:

$$|\boldsymbol{x} - \boldsymbol{x}_k| < \epsilon$$

4. Controller Design

As aforementioned, our mobile manipulator consists of a UR5 robotic arm and a Mecanum wheeled mobile platform. It is controlled by a control system, which is distributed on 2 control PCs. Fig. 6 shows the control architecture of the whole system. Details of the different controllers are described in the next sections.

4.1. *Control of the UR5 Manipulator*

The UR5 can be controlled at three different levels: The Graphical User Interface (GUI) Level, the Script Level and the C Application Programming Interface (C-API) Level[17]. *PolyScope* is the graphical user interface (GUI) for operating the robotic arm and for creating and executing robot programs. *URScript* is the robot programming language used to control the robot at the Script Level. This programs can be saved directly on the robot controller or commands can be sent via TCP/IP socket to the robot.

User programs that uses the C-API are executed on the UR controller and interact directly at the joint level with a cycle time of 8 ms. At this level, the UR controller can be supplied by either joint velocities or a combination of joint positions, joint velocities and joint accelerations. In our mobile manipulator, we use the C-API and control the UR5 at joint level. At this level, the synchronization of the platform movements with the arm movements is feasible in real-time. This enables trajectory planning for the whole system in Cartesian space and controlling the movements of platform and arm synchronously at joint level, while avoiding obstacles in real time, without leaving the trajectory.

The structure of the controllers with its interfaces is shown in Fig. 6. The robotic arm is controlled by the UR controller PC at joint level. The joint positions of the arm are generated by the path interpolation module in the system controller PC and streamed every 8 ms over a TCP/IP socket and Ethernet to the UR controller PC. The UR controller PC controls the joints and transfers the control signals over an interface module to the motors.

4.2. *Control of the Mecanum Wheeled Mobile Platform*

The controller of the Mecanum wheeled mobile platform is integrated in the system controller PC (platform control in Fig. 6). The internal structure of this controller is shown in Fig. 8. The path interpolation generates a stream of poses $\boldsymbol{x}_\mathrm{p} = (x_\mathrm{p}, y_\mathrm{p}, \theta_\mathrm{p})^\mathrm{T}$ and its derivatives $\dot{\boldsymbol{x}}_\mathrm{p}, \ddot{\boldsymbol{x}}_\mathrm{p}$ as input for the controller. The structure of the controller is derived from a classical cascade structure. It consists of a position controller, which controls the pose of the platform in world frame, a wheel controller, which controls the velocity of the wheels and a coupling controller, which controls the kinematic motion constraints.

The position controller obtains the actual pose of the platform by odometry and generates the velocities of the platform in the world frame plus

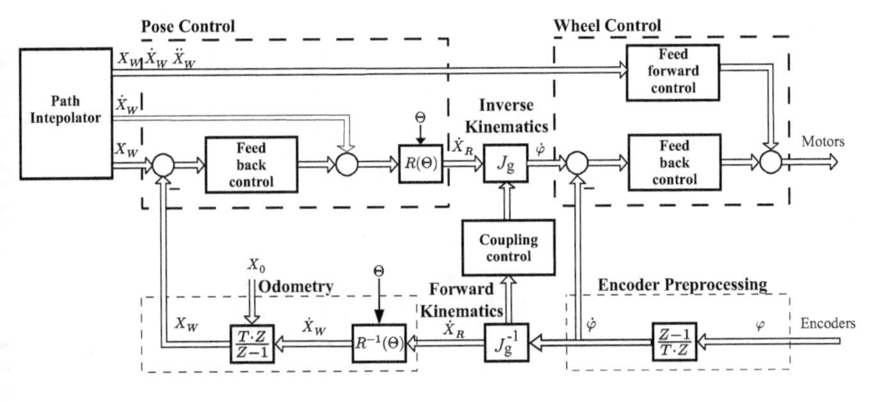

Figure 8. Controller structure of Mecanum wheeled platform

feed forwarding the velocities from the path interpolation. Details of the odometry for Mecanum wheeled platforms can be found in [18]. These velocities are transformed into the robot frame and then into reference velocities for the wheel controller. These reference velocities are calculated using the inverse kinematics of the platform (17) and therefore meet the kinematic constraints at any time. Owing to parameter differences in the control loops of the wheels or unbalanced loads, the actual wheel velocities may violate the kinematic motion constraints. These violations lead to additional wheel slippage.

Aim of the coupling controller is to change the reference velocities of the wheel controllers in such a way that the actual velocities meet the motion constraints [19]. The wheel controller controls the velocity of the wheels by feed back control using the wheel encoders plus optional calculated torque feed forward control.

Fig. 9 shows the software architecture of the navigation stack. The bottom of the navigation stack builds a hardware abstraction layer (HAL), which hides technical details of the communications with the robot [20]. The navigation stack contains 3 control loops: The pose control loop, the coupling controller at the level of wheel kinematics and the underlying wheel control loop. The localization layer obtains sensor data from HAL and estimates the pose of the platform using this data and odometry [18,21]. The top of the stack builds the motion planner for the platform.

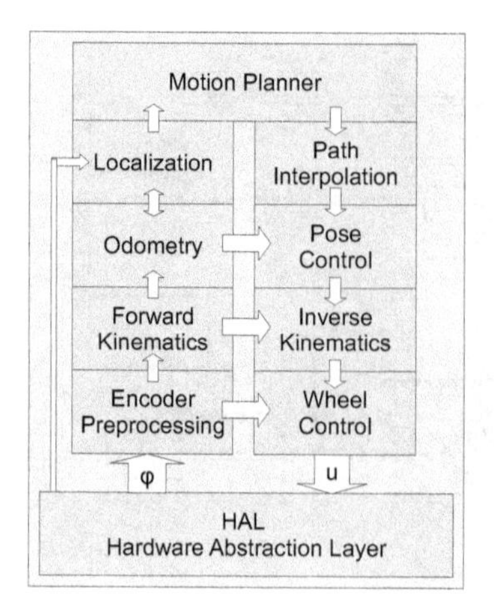

Figure 9. Software architecture of the navigation stack for controlling the mobile platform

5. Conclusions

In this paper, we have developed a compact and easy applicable kinematic model of an omnidirectional mobile manipulator. The mobile manipulator consists of a manipulator with 6 DoF and an over-actuated Mecanum wheeled mobile platform, which provides additional 3 DoF. The kinematic model includes the kinematic motion constraints of the mobile platform as well as the over-determined motion model of the whole mobile manipulator.

The kinematic model of the omnidirectional mobile manipulator is described by 9 DoF in the configuration space, which provide 6 DoF in Cartesian space. Therefore, the inverse kinematics are solved with a numerical approach using a Jacobi matrix. The proposed kinematic model is valid for all serial manipulators with 6 DoF, mounted on an omnidirectional mobile platform driven by $n \geq 4$ Mecanum wheels.

Acknowledgments

This work was supported by the Ministry of Culture and Science of the German State of North Rhine-Westphalia (grant number 005-1703-0008).

Bibliography

1. C. Röhrig and A. Jochheim, The virtual lab for controlling real experiments via Internet, in *Proceedings of the 11th IEEE International*

Symposium on Computer-Aided Control System Design, (Hawaii, USA, 1999).
2. R. Bischoff, U. Huggenberger and E. Prassler, KUKA youbot - a mobile manipulator for research and education, in *2011 IEEE International Conference on Robotics and Automation*, (IEEE, May 2011).
3. C.-H. King, T. L. Chen, A. Jain and C. C. Kemp, Towards an assistive robot that autonomously performs bed baths for patient hygiene, in *Intelligent Robots and Systems (IROS), 2010 IEEE/RSJ International Conference on*, (IEEE, 2010).
4. Y. Jiang, M. Lim, C. Zheng and A. Saxena, Learning to place new objects in a scene, *The International Journal of Robotics Research* **31**, 1021 (2012).
5. E. Dean-Leon, B. Pierce, F. Bergner, P. Mittendorfer, K. Ramirez-Amaro, W. Burger and G. Cheng, TOMM: Tactile omnidirectional mobile manipulator, in *Robotics and Automation (ICRA), 2017 IEEE International Conference on*, (IEEE, 2017).
6. M. Hvilshøj, S. Bøgh, O. Skov Nielsen and O. Madsen, Autonomous industrial mobile manipulation (AIMM): past, present and future, *Industrial Robot: An International Journal* **39**, 120 (2012).
7. C. Sprunk, B. Lau, P. Pfaff and W. Burgard, An accurate and efficient navigation system for omnidirectional robots in industrial environments, *Autonomous Robots* **41**, 473 (2017).
8. D. Wahrmann, A.-C. Hildebrandt, C. Schuetz, R. Wittmann and D. Rixen, An autonomous and flexible robotic framework for logistics applications, *Journal of Intelligent & Robotic Systems* **93**, 1 (2017).
9. C. Wurll, T. Fritz, Y. Hermann and D. Hollnaicher, Production logistics with mobile robots, in *ISR 2018; 50th International Symposium on Robotics*, (VDE, 2018).
10. D. Pavlichenko, G. M. García, S. Koo and S. Behnke, Kittingbot: A mobile manipulation robot for collaborative kitting in automotive logistics, in *Intelligent Autonomous Systems 15*, eds. M. Strand, R. Dillmann, E. Menegatti and S. Ghidoni (Springer International Publishing, Cham, 2019).
11. P. F. Muir and C. P. Neuman, Kinematic modeling for feedback control of an omnidirectional wheeled mobile robot, in *Robotics and Automation. Proceedings. 1987 IEEE International Conference on*, (IEEE, 1987).
12. C. Röhrig and D. Heß, OmniMan: An omnidirectional mobile manipulator for human-robot collaboration, in *Lecture Notes in Engineering*

and Computer Science: Proceedings of The International MultiConference of Engineers and Computer Scientists 2019, (Hong Kong, 2019).
13. G. Campion, G. Bastin and B. Dandrea-Novel, Structural properties and classification of kinematic and dynamic models of wheeled mobile robots, *IEEE Transactions on Robotics and Automation* **12**, 47 (1996).
14. Universal Robots A/S, Odense, Denmark, *UR5 Technical Specifications*, (2015). `http://www.universal-robots.com/media/50588/ur5_en.pdf`.
15. K. P. Hawkins, *Analytic Inverse Kinematics for the Universal Robots UR-5/UR-10 Arms*, tech. rep., Georgia Institute of Technology (2013), `https://smartech.gatech.edu/bitstream/handle/1853/50782/ur_kin_tech_report_1.pdf`.
16. F. Künemund, D. Heß and C. Röhrig, Energy efficient kinodynamic motion planning for holonomic AGVs in industrial applications using state lattices, in *Proceedings of the 47th International Symposium on Robotics (ISR 2016)*, (Munich, Germany, 2016).
17. Universal Robots A/S, Odense, Denmark, *The URScript Programming Language, Version 3.1* (January, 2015).
18. C. Röhrig, Heß and F. Künemund, RFID-based localization of mobile robots using the received signal strength indicator of detected tags, *Engineering Letters* **24**, 338 (August 2016).
19. C. Röhrig, D. Heß and F. Künemund, Motion controller design for a mecanum wheeled mobile manipulator, in *Proceedings of the IEEE Conference on Control Technology and Applications (CCTA 2017)*, (Kohala Coast, Hawai'i, USA, 2017).
20. D. Heß, F. Künemund and C. Röhrig, Linux based control framework for mecanum based omnidirectional automated guided vehicles, in *Lecture Notes in Engineering and Computer Science: Proceedings of The World Congress on Engineering and Computer Science 2013, WCECS 2013*, (San Francisco, USA, 2013).
21. C. Röhrig, C. Kirsch, J. Lategahn, M. Müller and L. Telle, Localization of Autonomous Mobile Robots in a Cellular Transport System, *Engineering Letters* **20**, 148 (June 2012).

A Simple Baseline Correction Method for Raman Spectra

Yung-Sheng Chen* and Yu-Ching Hsu

Department of Electrical Engineering, Yuan Ze University,
Chungli, Taoyuan 320, Taiwan, ROC
**E-mail: eeyschen@saturn.yzu.edu.tw*

Baseline shifts caused by a variety of factors depending on the type of spectroscopy are common degradation problems in Raman spectra. An effective and efficient method is presented for the baseline correction of Raman spectra. A kernel smoothing function is developed for the estimation of baseline, and a negative signal filter is designed for the removal of negative impulse responses caused from the spectroscopy. Results on the computer and instrument implementation confirm the feasibility of the proposed method.

Keywords: Raman spectra; Spectrometers; Spectroscopic instrumentation; Spectroscopy.

1. Introduction

In applied sciences, a lot of natural phenomena can be observed and measured based on a proper instrument for further analyzing and studying. For example, in order to read a color-band resistor (which is an often-used electronic component), except for the use of multimeter measurement or human observation and calculation, it can also be read by a computer vision based method.[1–3] For the multimeter, it is also a quite significant instrument for sensing or measuring electronic parameters (e.g., voltage, current, resistance) and is indispensable to the area of science and technology. However, reading an analog multimeter usually relies on human eyes and has two obvious problems, i.e., inefficiency and easy fatigue, while a long time of reading the analog multimeter is needed. This conducts us to develop a computer vision approach for automatically reading the analog multimeter in recent.[4,5] From such a research trend of instrument, the signal processing method to the instrument of Raman spectroscopy is similar to the reading technique (such as the computer vision method) to the analog multimeter. An effective and efficient baseline correction algorithm for Raman spectra is thus addressed in this paper.

Raman spectroscopy is a well-established but significant technique for revealing the molecular fingerprints based on its vibrational information. Due to the label-free and non-invasive property as well as convenient application, Raman spectroscopy has been widely applied for material analysis,[6] biological investigation,[7] medical diagnosis,[8] and so on. Baseline is a common degradation problem in Raman spectra and usually results from the spurious background signal or instrument fluctuation. The problem of baseline will severely result in the difficulty of quantitative or qualitative analysis of Raman spectra. To overcome such a problem, baseline correction is a necessary step before performing the Raman spectra analysis and usually classified into three main types of methods, i.e., smoothing spline fitting, wavelet decomposition and integration, and least squares error modelling (or divided into physically and mathematically motivated approaches). The merits and disadvantages of those methods have been investigated and can be found in the related literatures.[6,7,9] A well-known commercial software tool, called Origin/OriginPro developed by OriginLab Corporation,[10] have been widely used in the related application fields, where the peak analysis is one of the major functions in this product and includes the baseline correction function being focused in our study. In recent, due to the progress of micro Raman spectroscopy, e.g., Micro Raman Identify (MRI) spectrometer,[11] users usually need a friendly interface to fast correct the baseline with only a few controllable parameters. To achieve such a goal, an effective and efficient baseline correction algorithm is presented in this paper. The partial work of this study has been presented in the IMECS 2019.[12]

The rest of this paper is organized as follows. In Section 2, an illustration of base line estimation and correction is given and the base line modeling is introduced at first. Then a kernel smooth function and a so-called negative signal removal are developed to construct our baseline correction algorithm. Section 3 shows the experimental results, where the behaviour of the parameters is also discussed. The conclusion and future work of this paper is finally drawn in Section 4.

2. Proposed Approach

All the samples of Raman spectra used in this study are from the MRI spectrometer.[11] Two samples are illustrated in Fig. 1, where the baseline (red colour) is estimated by the proposed method. Note that the negative impulse responses result from the MRI spectrometer can also be filtered out by the proposed method as illustrated in Fig. 1(b). Their final baseline correction results are shown respectively in Fig. 2. The details of the

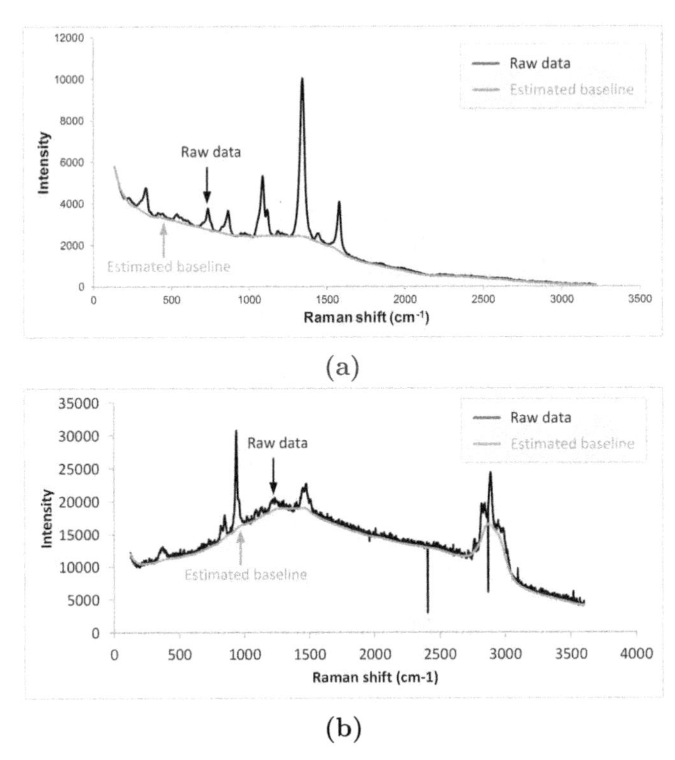

Fig. 1. Illustrations of the measured Raman spectrum and the baseline (red colour) estimated by the proposed algorithm. (a) Sample-1: Measured Raman spectrum including the baseline. (b) Sample-2: Measured Raman spectrum including not only the baseline but also the negative impulse responses.

presented algorithm will be described as follows.

2.1. *Baseline removal modeling*

Let f_{raw} be the raw observed Raman spectrum having N data points. Generally, the raw data f_{raw} is composed of the true signal (s), the baseline (b), as well as the noise (n). It can thus be modelled as

$$f_{\text{raw}} = s + b + n \tag{1}$$

Based on the quality of nowadays technology, the noise term n can be ignored since it has a little effect compared to the baseline. Therefore, the signal s can be estimated by subtracting the baseline b from the raw data f_{raw}. In other words, the estimation of baseline is the main task for extracting the signal s. Let f_{baseline} be the estimated baseline. Then the

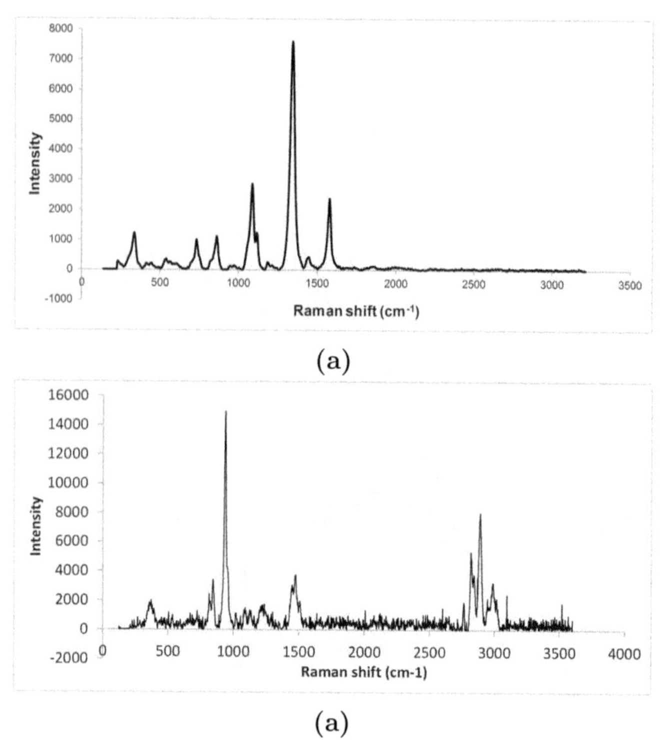

(a)

Fig. 2. Results of baseline correction by the proposed algorithm. (a) Baseline correction result for the Raman spectrum given in Fig. 1(a). (b) Baseline correction result for the Raman spectrum given in Fig. 1(b).

correct signal f_{correct} can be expressed as

$$f_{\text{correct}} = f_{\text{raw}} - f_{\text{baseline}} \tag{2}$$

2.2. *Smooth function*

From the characteristic of Raman spectra, the baseline can be regarded as a stable component from the human visual inspection, which is used to carry the signal. Therefore, the key function of estimating the baseline is first developed in this study for finding the local minima of the raw data and smoothing them as the **SMOOTH** function depicted in Fig. 3(a). Here f_{in} is the input signal, whereas the $\overline{f}_{\text{min}}$ and $\overline{f}_{\text{avg}}$ are the resultant local minima and averaged information. Assume the i-th data point is to be processed, the sliding window is defined as $[i - W, i + W]$, where W is a positive integer and controls the window size. The functional blocks

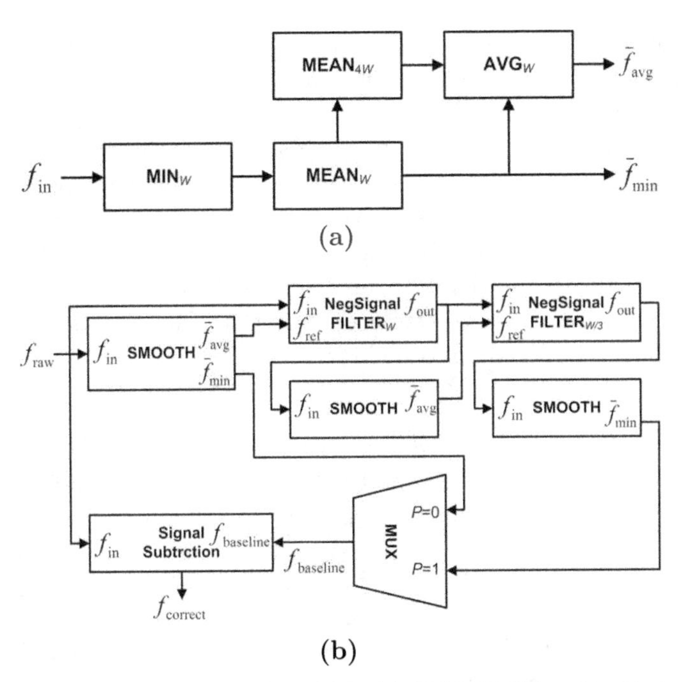

Fig. 3. Signal flow of the proposed method. (a) **SMOOTH** function block diagram with **MIN**, **MEAN** and **AVG** operations for finding useful local stationary information. (b) Flowchart of removing the negative impulse responses, extracting the baseline, and obtaining the baseline correction result.

(**MIN**, **MEAN**, and **AVG**) shown in Fig. 3(a) are formulated respectively as follows.

$$\mathrm{MIN}_W(i) = \min_{\forall j \in [i-W, i+W]} f_{\mathrm{in}}(j) \tag{3}$$

$$\mathrm{MEAN}_W(i) = \left(\sum_{\forall j \in [i-W, i+W]} \mathrm{MIN}_W(j) \right) / (2W+1) \tag{4}$$

$$\mathrm{MEAN}_{4W}(i) = \left(\sum_{\forall j \in [i-4W, i+4W]} \mathrm{MEAN}_W(j) \right) / (8W+1) \tag{5}$$

$$\mathrm{AVG}_W(i) = \left(\mathrm{MEAN}_W(i) + \mathrm{MEAN}_{4W}(i) \right) / 2 \tag{6}$$

Here $\mathbf{MIN}_W$ computes the fundamental local minima for the input signal and after the processing of $\mathbf{MEAN}_W$, the smoothed local minima $\overline{f}_{\min}$

can be obtained. However, in our study on the MRI spectrometer,[11] some negative impulse responses (possibly resulting from the sensor device or the integrated mechanism) may occur during the signal acquisition as the Sample-2 shown in Fig. 1(b). To filter out such an unwanted negative signal, it is reasonable to provide a more smoother local minimum as a reference for comparison. A more wider $\mathbf{MEAN}_{4W}$ function is thus performed on the resultant signal from $\mathbf{MEAN}_W$ and then combined with $\mathbf{MEAN}_W$ by $\mathbf{AVG}_W$ to obtain the useful reference signal $\overline{f}_{\text{avg}}$. As long as the mean input signal is less than the reference, it can be replaced by the mean signal, otherwise the original input signal is remained.

2.3. *Negative signal removal*

The function block of $(\mathbf{NegSignal\ Filter})_W$ in Fig. 3(b) performs such a negative signal removal, which can be formulated as below.

$$f_{\text{out}}(i) = \begin{cases} \overline{f}_{\text{ref}}(i), \ \overline{f}_{\text{in}}(i) < \overline{f}_{\text{ref}}(i) \\ f_{\text{in}}(i), \ \ \text{otherwise} \end{cases} \tag{7}$$

Here $\overline{f}_{\text{in}}$ represents the resultant mean signal from the input signal f_{in}, and can be computed as the formula in Eq. (4) with W windows size. $\overline{f}_{\text{ref}}$ is obtained by the same way. Since the negative impulse responses may be large or small as illustrated in Fig. 1(b), $(\mathbf{NegSignal\ Filter})_W$ cascading (**SMOOTH**) and $(\mathbf{NegSignal\ Filter})_{W/3}$ cascading (**SMOOTH**) are adopted respectively for the removal of large and small negative signal as depicted in Fig. 3(b). It can be regarded as a two-stage negative signal filter. Here the **SMOOTH** extracts continuously the useful $\overline{f}_{\text{min}}$ and $\overline{f}_{\text{avg}}$ from the filtered signal obtained by the **NegSignal Filter**. In this case, the final baseline is estimated and output $\overline{f}_{\text{min}}$ located at the middle right of the signal flow given in Fig. 3(b). Note here that if there is not any negative impulse responses such a two-stage negative signal filter would not be performed, the baseline can be directly output from the $\overline{f}_{\text{min}}$ of the first **SMOOTH** function block.

2.4. *Baseline correction algorithm*

The whole signal flow of the proposed algorithm is depicted in Fig. 3(b). Let P be a binary parameter to control a multiplexer (**MUX**), which outputs the estimate baseline f_{baseline} that comes from the two-stage negative signal filter ($P = 1$) or the first **SMOOTH** function block ($P = 0$). The baseline correction result f_{correct} can be further obtained by the **Signal**

Subtraction function block. For the given Raman spectra shown in Fig. 1, the estimated baselines (red colour) and their baseline correction results shown in Fig. 2 demonstrate the feasibility of the proposed algorithm.

Table 1. Performances of the five samples from the Micro Raman Identify (MRI) spectrometer[11] using the proposed algorithm.

Samples	**data size**	W	P	**execution time (ms)**
Sample-1	1559	19	0	5
Sample-2	1979	15	1	15
Sample-3	1010	15	0	3
Sample-4	1014	15	0	3
Sample-5	1973	31	1	23

3. Result and Discussion

The algorithm is implemented in Microsoft® Visual Studio® C++ 2013 and run on a laptop with Intel® Core™ 2 Duo 1.8 GHz CPU and 4G RAM. In addition to the baseline correction results of Sample-1 and Sample-2 shown in Fig. 2, the results of other three samples are also given in Fig. 4 for further evaluations. Fig. 4(a) shows our method can estimate effectively the baseline for a background spectrum (Sample-3). Fig. 4(b) shows the result of a material measured under such a background (Sample-4). Fig. 4(c) shows the baseline estimation result for a spectrum with some negative impulse responses corrupted, and its baseline correction result is shown in Fig. 4(d) (Sample-5). The performances reported in Table 1 are evaluated by using data size, W, P, as well as execution time (with milliseconds or ms), where parameters W and P can be selected by user. It can be observed that the execution time of the presented method is positively related to the data size and the window size W.

Before concluding this paper, the behaviours of W and P are briefly discussed as follows. Considering the Sample-1, the results of baseline estimation using parameter (a) $W = 7$, (b) $W = 9$, and (c) $W = 31$ are shown in Fig. 5. By comparing the result using $W = 19$ shown in Fig. 1(a), it can be observed that the small W (see Fig. 5(a) and 5(b)) will result in an under-fitting, whereas the large W (see Fig. 5(c)) will result in an over-fitting. Since there are possibly a lot of various Raman spectra, it is reasonable to select a proper W by means of heuristic manner. Considering

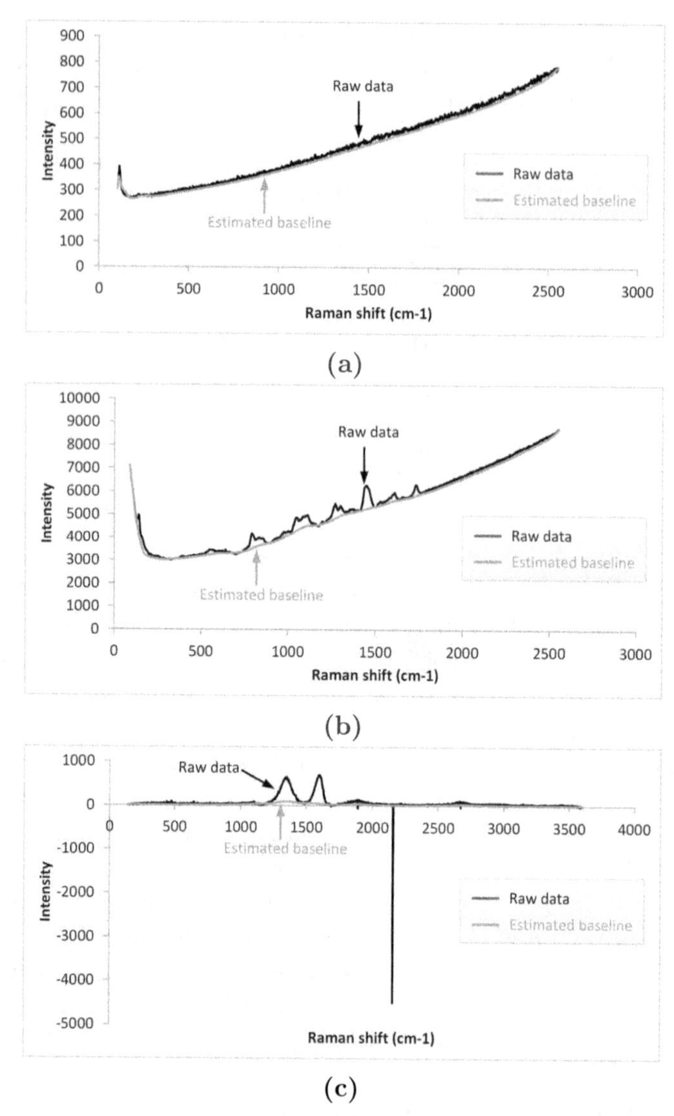

Fig. 4. Results of the other samples for baseline correction. (a) The estimated baseline fits well for a background spectrum. (b) Result of a material measured under such a background in (a). (c) The estimated baseline for a spectrum with some negative impulse responses corrupted. (d) The final baseline correction result for the spectrum in (c).

the Sample-2, with the same $W = 15$ and set $P = 0$, it can be observed that the baseline estimation is influenced with the negative impulse signal as the result shown in Fig. 6 by comparing the result given in Fig. 1(b),

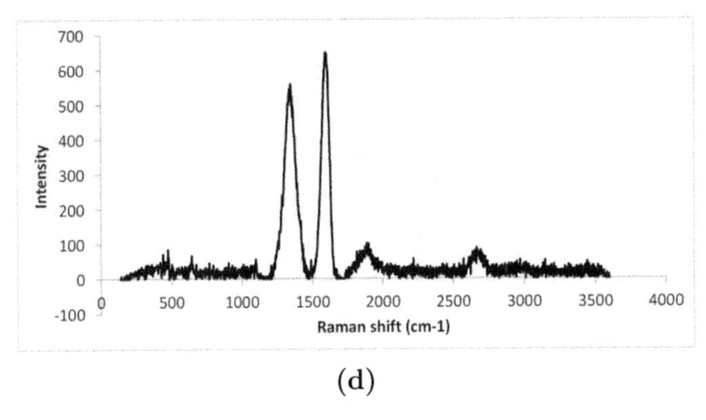

(d)

Fig. 4. (Continued.)

where the negative signal is removed by setting $P = 1$.

Based on our experiments, the effectiveness and efficiency of the proposed baseline correction algorithm for Raman spectra have been confirmed. As a result, in accordance with the desired application of micro Raman spectroscopy, e.g., Micro Raman Identify (MRI) spectrometer,[11] the goal of providing users a friendly interface to fast correct the baseline with only a few controllable parameters has been achieved by the presenting approach.

4. Conclusion and Future Work

In the field of Raman spectrometer, it is of great importance to correct the baseline before performing the quantitative and qualitative analysis for a raw Raman spectrum. The presented algorithm can effectively not only correct the baseline but also remove the negative impulse responses. The execution time with milliseconds performs the efficiency of the proposed method. This algorithm has been implemented in the instrument software of Micro Raman Identify (MRI) spectrometer[11] and thus confirms its feasibility in the engineering and laboratory application. Because the peak fitting or peak decomposition is also a significant function in the field of peak analysis, it could be a good topic along this study direction and regarded as our future work.

Acknowledgment

This work was supported in part by the Protrustech (Taiwan), which is a professional instrument company, working on Raman spectroscopy, Photoluminescence, Photo-reflectance, and so on.

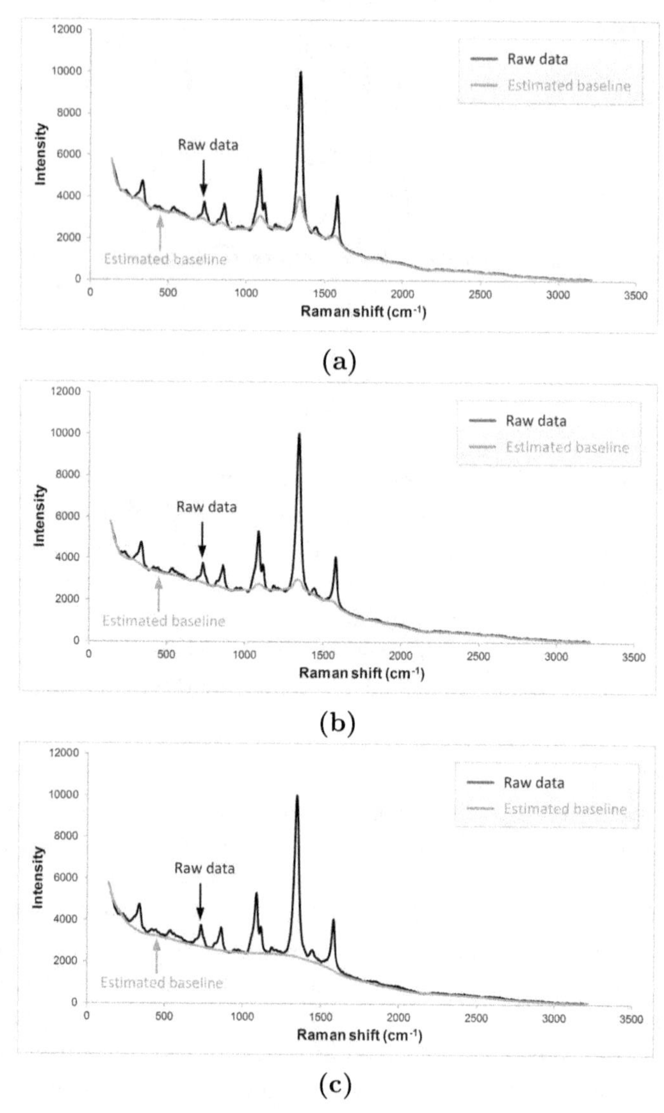

Fig. 5. Results for Sample-1 with the parameter: (a) $W = 7$ (sever under-fitting), (b) $W = 11$ (little under-fitting), (c) $W = 31$ (over-fitting).

References

1. Y.-S. Chen and J.-Y. Wang, Reading resistor based on image processing, in *Proc. IEEE Int. Conference on Machine Learning and Cybernetics, ICMLC 2015*, (Guangzhou, China, 2015, pp566–571).

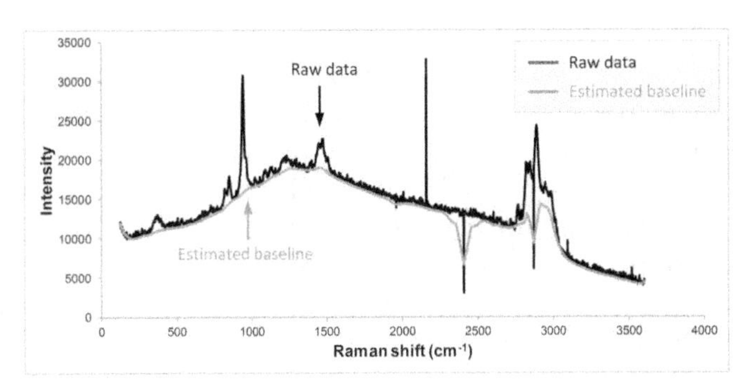

Fig. 6. Result for Sample-2 with the parameter $P = 0$, where the negative signal removal is not performed.

2. Y.-S. Chen and J.-Y. Wang, Implementation of cost-effective diffuse light source mechanism to reduce specular reflection and halo effects for resistor-image processing, in *Proc. of SPIE Optical Engineering + Applications: Applications of Digital Image Processing XXXVIII*, (San Diego, CA, USA, 2015, pp959928-1-14).
3. Y.-S. Chen and J.-Y. Wang, Computer vision on color-band resistor and its cost-effective diffuse light source design, *Journal of Electronic Imaging* **25**, 061409 (2016).
4. Y.-S. Chen and J.-Y. Wang, A novel approach of reading analog multimeter based on computer vision, in *Proc. IEEE Int. Conference on Applied System Innovation, ICASI 2018*, (Chiba, Tokyo, Japan, 2018, pp758–761).
5. Y.-S. Chen and J.-Y. Wang, Computer vision-based approach for reading analog multimeter, *Applied Sciences* **8**, 1268 (2018).
6. M. Koch, C. Suhr, B. Roth and M. Meinhardt-Wollweber, Iterative morphological and mollifier-based baseline correction for raman spectra, *Journal of Raman Spectroscopy* **48**, 336 (2017).
7. S. Guo, T. Bocklitz and J. Popp, Optimization of raman-spectrum baseline correction in biological application, *Analyst* **141**, 2396 (2016).
8. P. Kumar, T. Bhattacharjee, M. Pandey, A. Hole, A. Ingle and C. M. Krishna, Raman spectroscopy in experimental oral carcinogenesis: investigation of abnormal changes in control tissues, *Journal of Raman Spectroscopy* **47**, 1318 (2016).
9. J. Liu, J. Sun, X. Huang, G. Li and B. Liu, Goldindec: a novel algorithm for raman spectrum baseline correction, *Applied Spectroscopy* **69**, 834 (2015).

10. Peak Analisys in the software of Origin or OriginPro, OriginLab Corporation, USA, `https://www.originlab.com/`.
11. Raman Spectroscopy, Protrustech CO., LTD, Taiwan, `https://www.protrustech.com/mri.html`.
12. Y.-S. Chen and Y.-C. Hsu, Effective and efficient baseline correction algorithm for raman spectra, in *Lecture Notes in Engineering and Computer Science: Proceedings of The International MultiConference of Engineers and Computer Scientists 2019, IMECS 2019*, (Hong Kong, 13-15 March, 2019, pp295–298).

Biquadratic Allpass Phase-Compensating System Design Utilizing Bilinear Error

Tian-Bo Deng

Department of Information Science
Faculty of Science, Toho University
Miyama 2-2-1, Funabashi, Chiba 274-8510, Japan

deng@is.sci.toho-u.ac.jp

A new method is proposed for the design of a biquadratic (biquad) allpass phase-compensating system whose stability-margin satisfies an arbitrarily preset stability criterion. The design technique utilizes both the generalized stability-triangle (GST) and the bilinear (BL) phase-error function of the allpass biquad system, which are derived by the author. Based on the GST condition, it is shown that the original coefficients of the allpass biquad system can be converted into the functions of other two new variables, while the two variables have no limits (bounds) on their values to satisfy the GST condition. That is, arbitray values of the two new variables can meet the GST condition. Based on the above variable conversions, the design technique further employs a nonlinear optimization method to find the optimum values of the two variables to approximate a prescribed ideal phase. Thanks to the variable conversitions, the resulting values of the two variables never violate the GST condition. As a result, the resulting biquad allpas system not only can always satisfy a prescribed stability margin, but also can best approximate a given ideal phase in the minimax sense. To demonstrate the guaranteed stability margin as well as the system accuracy, a biquad system design example is included.

Keywords: Bilinear (BL) error function; biquadratic (biquad) system; stability; generalized stability-triangle (GST); stability margin.

1. Introduction

Digital phase-compensating systems are the basic circuits that are required in the phase-compensations of signal processing sytems and digital communication systems when the systems have phase distortions. Since the phase distortions will cause the undesirable distortions of signal waveforms, it is mandatory to calibrate such phase distortions for preserving the signal waveforms. As is well known, the design of a finite-impulse-response phase system needs to approximate both the amplitude and phase characteristics[1-6], while the design of a recursive allpass one only needs to elaborate its phase approximation[7-11]. However, it is also well known that the latter is a recursive system and its stability needs to be ensured. Moreover, instead of just guaranteeing its stability, it is also preferred to let a recursive system have an extra stability margin because some coefficient-value perturbations such as coefficient-quantization errors may cause instability. In[12], a new stability triangle called generalized stability-triangle (GST) is derived, and this GST allows the system designer to add a preset margin to the system stability during the system design[13,14].

This paper formalizes an alternative way for the design of a biquadratic (biquad) allpass phase system whose phase approximates a prescribed phase characteristic. The mathematic model of the biquad system is the second-order rational function, and it has two independent system coefficients. Due to the nonlinearity of its phase response, i.e., its phase is nonlinearly related with the two system coefficients, this forces the phase-approximation problem to be a nonlinear minimization problem. To express the phase as an explicit function of the two coefficients, we give the detailed derivation of an approximate phase error whose numerator and denominator are linear functions of the two system coefficients. For simplicity, this phase error is referred to as bilinear (BL) error. This BL error can be viewed as a special case of the high-order system[11]. Then, the BL error is employed together with the GST condition in the design of a biquad allpass system, which is a nonlinear optimization problem. It is demonstrated through a biquad allpass system design example that the obtained biquad allpass system not only satisfies a preset stability margin, but also has small design errors.

2. Bilinear Error

Suppose that we are given the desired phase $\Theta_d(\omega)$ and it is approximated uing the allpass biquad-system function

$$\begin{aligned} H(z) &= \frac{c_2 + c_1 z^{-1} + z^{-2}}{1 + c_1 z^{-1} + c_2 z^{-2}} \\ &= z^{-2} \cdot \frac{1 + c_1 z + c_2 z^2}{1 + c_1 z^{-1} + c_2 z^{-2}} \\ &= z^{-2} \cdot \frac{C(z^{-1})}{C(z)} \end{aligned} \tag{1}$$

where

$$C(z) = 1 + c_1 z^{-1} + c_2 z^{-2}$$

denotes the denominator of the biquad-system function, and this denominator is the second-order polynomial in z^{-1}. Similarly,

$$C(z^{-1}) = 1 + c_1 z + c_2 z^2$$

is part of the numerator of $H(z)$. Obviously, both $C(z)$ and $C(z^{-1})$ have the same coefficients $\{c_1, c_2\}$. Fig. 1 illustrates the block-diagram of this biquad system.

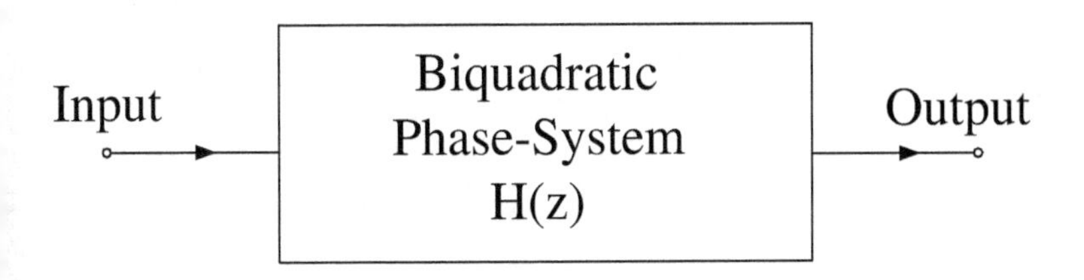

Fig. 1. Biquad phase-compensating system.

To formalize the approximation problem, we give the detailed derivation of the BL phase error as follows. This BL error is to be minimized in the design.

By writing the frequency response of the quadratic polynomial $C(z)$ as

$$\begin{aligned} C(\omega) &= 1 + c_1 e^{-j\omega} + c_2 e^{-j2\omega} \\ &= \sum_{k=0}^{2} c_k e^{-jk\omega} \end{aligned}$$

with $c_0 = 1$, one can readily obtain its complex-conjugate expression

$$C^*(\omega) = \sum_{k=0}^{2} c_k e^{jk\omega}.$$

Using $C(\omega)$ and $C^*(\omega)$, one yields the frequency response of the biquad-system $H(z)$ as

$$H(\omega) = e^{-j2\omega} \cdot \frac{C^*(\omega)}{C(\omega)}.$$

Let the phase of $C^*(\omega)$ be $\beta(\omega)$, i.e.,

$$\beta(\omega) = \angle C^*(\omega). \tag{2}$$

The overall phase of $H(\omega)$ can be expressed as

$$\begin{aligned}\Theta(\omega) &= 2\beta(\omega) - 2\omega \\ &= 2\left[\beta(\omega) - \omega\right].\end{aligned}$$

In the extreme case where the approximation error equals zero, we have

$$\Theta(\omega) = \Theta_d(\omega)$$

i.e.,

$$2\left[\beta(\omega) - \omega\right] = \Theta_d(\omega).$$

Equivalently,

$$\beta(\omega) = \omega + \Theta_d(\omega)/2.$$

This means that the ideal phase for $\beta(\omega)$ should be

$$\beta_d(\omega) = \omega + \Theta_d(\omega)/2. \tag{3}$$

To simplify the notations, we denote the error between $\beta(\omega)$ and $\beta_d(\omega)$ by

$$e_\beta(\omega) = \beta(\omega) - \beta_d(\omega). \tag{4}$$

Consequently, the overall phase error can be expressed as

$$\begin{aligned}e_\Theta(\omega) &= \Theta(\omega) - \Theta_d(\omega) \\ &= 2[\beta(\omega) - \omega] - \Theta_d(\omega) \\ &= 2\left[\beta(\omega) - \left(\omega + \frac{\Theta_d(\omega)}{2}\right)\right] \\ &= 2e_\beta(\omega).\end{aligned} \tag{5}$$

The relation in (5) implies that it suffices to minimize $|e_\beta(\omega)|$ in order to minimize $|e_\Theta(\omega)|$. To minimize $|e_\beta(\omega)|$, we first derive a cost function called BL phase error as follows.

Let

$$C^*(\omega) = C_{\mathrm{R}} + jC_{\mathrm{I}}$$

with

$$\begin{aligned}C_{\mathrm{R}} &= \sum_{k=0}^{2} c_k \cos(k\omega) \\ C_{\mathrm{I}} &= \sum_{k=0}^{2} c_k \sin(k\omega).\end{aligned}$$

Hence, $e_\beta(\omega)$ can be further detailed as

$$\begin{aligned} e_\beta(\omega) &= \beta(\omega) - \beta_d(\omega) \\ &= \tan^{-1}\frac{C_\mathrm{I}}{C_\mathrm{R}} - \beta_d(\omega). \end{aligned}$$

When $e_\beta(\omega)$ is small enough, we can use $\tan(e_\beta(\omega))$ to approximate $e_\beta(\omega)$ as

$$\begin{aligned} e_\beta(\omega) &\approx \tan(e_\beta(\omega)) \\ &= \tan(\beta(\omega) - \beta_d(\omega)) \\ &= \frac{\tan\beta - \tan\beta_d}{1 + \tan\beta \cdot \tan\beta_d} \\ &= \frac{C_\mathrm{I}/C_\mathrm{R} - \sin\beta_d/\cos\beta_d}{1 + C_\mathrm{I}/C_\mathrm{R} \cdot \sin\beta_d/\cos\beta_d} \\ &= \frac{C_\mathrm{I}\cos\beta_d - C_\mathrm{R}\sin\beta_d}{C_\mathrm{R}\cos\beta_d + C_\mathrm{I}\sin\beta_d} \\ &= \frac{\sum_{k=0}^{2} c_k[\sin(k\omega)\cos(\beta_d) - \sin(\beta_d)\cos(k\omega)]}{\sum_{k=0}^{2} c_k[\cos(k\omega)\cos(\beta_d) + \sin(k\omega)\sin(\beta_d)} \\ &= \frac{\sum_{k=0}^{2} c_k \sin(k\omega - \beta_d)}{\sum_{k=0}^{2} c_k \cos(k\omega - \beta_d)} \end{aligned} \tag{6}$$

where $\beta_d(\omega)$ is shortened to β_d. Here, we let

$$\epsilon_\beta(\omega) = \frac{\sum_{k=0}^{2} c_k \sin(k\omega - \beta_d)}{\sum_{k=0}^{2} c_k \cos(k\omega - \beta_d)} = \frac{N(\omega)}{D(\omega)}. \tag{7}$$

In (7), it is clear that both the numerator

$$N(\omega) = \sum_{k=0}^{2} c_k \sin(k\omega - \beta_d)$$

and the denominator

$$D(\omega) = \sum_{k=0}^{2} c_k \cos(k\omega - \beta_d)$$

are linear functions of the system coefficients $\{c_1, c_2\}$. This is the reason why $\epsilon_\beta(\omega)$ is called bilinear (BL) error. The above derivation result can be viewed as a special case of the derivation in[11], and this BL expression explictly relates the phase error with the two independent coefficients $\{c_1, c_2\}$.

3. GST-Based Minimax Design

In Fig. 2, the dotted line is the conventional stability triangle (CST), and the solid line depicts the generalized stability-triangle (GST) derived in[12]. This section proposes a minimax design using the GST and shows that the derived GST enables the designer to design a second-order recursive system with a preset stability margin.

In Fig. 2, the proposed GST is embedded inside the CST, and when $\rho = 1$, the GST becomes the CST. Moreover, the interior of the GST and its three sides constitute a closed stable region. For a fixed value of ρ, if the

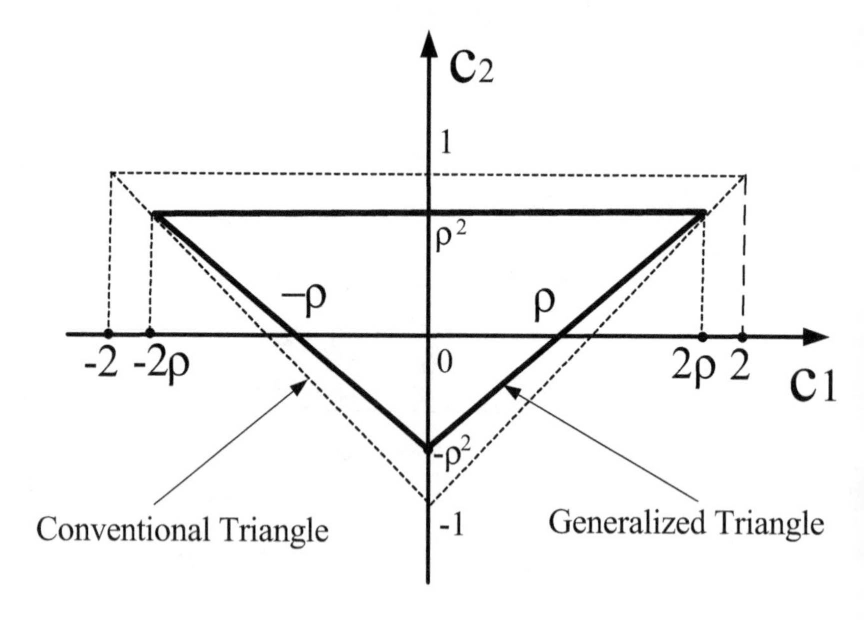

Fig. 2. The GST embedded inside the CST.

two coefficients $\{c_1, c_2\}$ of the biquad allpass system satisfy the condition

$$\begin{cases} |c_2| \leq \rho^2 \\ |c_1| \leq \rho + \rho^{-1} c_2 \end{cases} \tag{8}$$

the two poles p_1 and p_2 of the biquad transfer function in (1) meet the condition

$$|p_i| \leq \rho, \ i = 1, 2.$$

Therefore, the parameter ρ represents the upper bound on the radii of the two poles, and $0 < \rho < 1$. A small value of ρ implies that the two poles of the allpass system are far from the unit circle, and the recursive system has robust stability.

To employ the GST condition (8) in designing the biquad system, the biquad-system coefficients $\{c_2, c_1\}$ are transformed to the functions of two new variables $\{x_2, x_1\}$ as

$$\begin{cases} c_2 = U(x_2)\rho^2 \\ c_1 = U(x_1)(\rho + \rho^{-1} c_2) \\ \quad = U(x_1)[\rho + \rho U(x_2)] \\ \quad = \rho U(x_1)[1 + U(x_2)] \end{cases} \tag{9}$$

by using the function

$$U(x) \in [-1, 1].$$

Since the function $U(x)$ satisfies the condition

$$|U(x)| \leq 1$$

we call such a function *unity-bounded* function. It has been rigorously proved that whatever values of the new parameters $\{x_2, x_1\}$ always guarantee that the the GST condition in (8) is met[13]. Based on the transformations in (9), the BL error in (7) becomes

$$\epsilon_\beta(\omega) = \frac{\sin(-\beta_d) + \rho U(x_1)[1 + U(x_2)]\sin(\omega - \beta_d) + \rho^2 U(x_2)\sin(2\omega - \beta_d)}{\cos(-\beta_d) + \rho U(x_1)[1 + U(x_2)]\cos(\omega - \beta_d) + \rho^2 U(x_2)\cos(2\omega - \beta_d)}. \tag{10}$$

It is obvious that the BL error becomes a nonlinear function of the new variables $\{x_2, x_1\}$, and finding the optimum values of $\{x_2, x_1\}$ is the final objective of the minimax design.

To perform the minimax design, following steps are needed. First, the original system coefficients $\{c_2, c_1\}$ are expressed as (9) using the unity-bounded function as well as the two new variables $\{x_2, x_1\}$. Then, the variables $\{x_2, x_1\}$ are optimized via minimizing the peaked error

$$\Delta_\beta = \max\{|\epsilon_\beta(\omega)|,\ \omega \in [0, \pi]\}. \quad (11)$$

Minimizing the maximum of $|\epsilon_\beta(\omega)|$ means minimizing the worst-case deviation, so such a design is referred to as minimax design. This minimax design is a nonlinear mathematical programming problem, which is accomplished by using the nonlinear optimizer (*fminimax* in MATLAB). To optimize $\{x_2, x_1\}$, initial values of $\{x_2, x_1\}$ are first given, and then *fminimax* iteratively seeks for the optimum solutions of $\{x_2, x_1\}$. Once the optimum values of $\{x_2, x_1\}$ are obtained, we evaluate the performance (accuracy) of the resulting biquad-system by using the peaked error of the phase response

$$\Delta_\Theta = \max\{|e_\Theta(\omega)|,\ \omega \in [0, \pi]\}. \quad (12)$$

4. Biquad-System Design Example

To demonstrate the effectiveness of the proposed minimiax design approach using both the BL phase error and the GST, this section gives a biquad-system design example to show the design accuracy as well as the obtained stability margin.

Fig. 3 shows the ideal phase

$$\Theta_d(\omega) = -3\omega + \pi \sin^2(\omega/2) \quad (13)$$

that is to be approximated by using the proposed minimax design method[14].

First, the proposed design method needs a unity-bounded functon for the transformations in (9). As in[13,14], we use

$$U(x) = \sin(x)$$

as shown in Fig. 4 for transforming the coefficients (c_2, c_1) into the new variables (x_2, x_1) as

$$\begin{cases} c_2 = \rho^2 \sin(x_2) \\ c_1 = \rho \sin(x_1)[1 + \sin(x_2)]. \end{cases}$$

To execute the nonlinear optimizer *fminimax*, the initial values of

$$(x_2, x_1) = (0, 0)$$

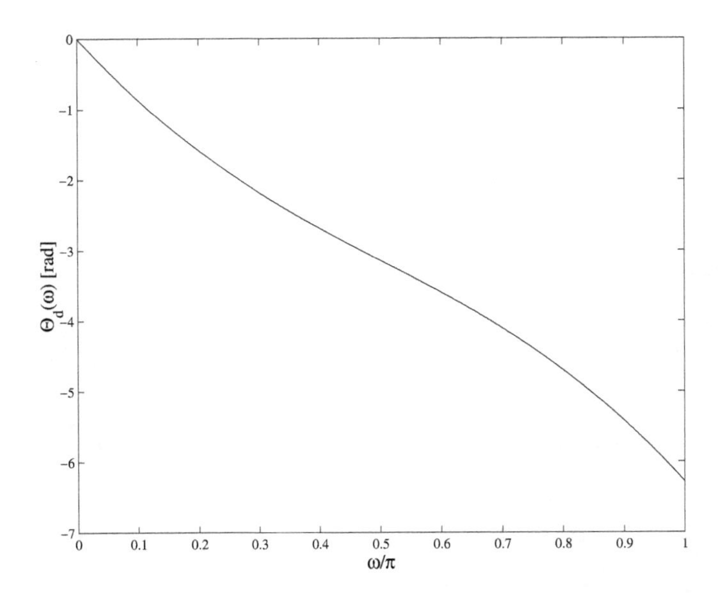

Fig. 3. Desired phase for the BL phase-error-based design.

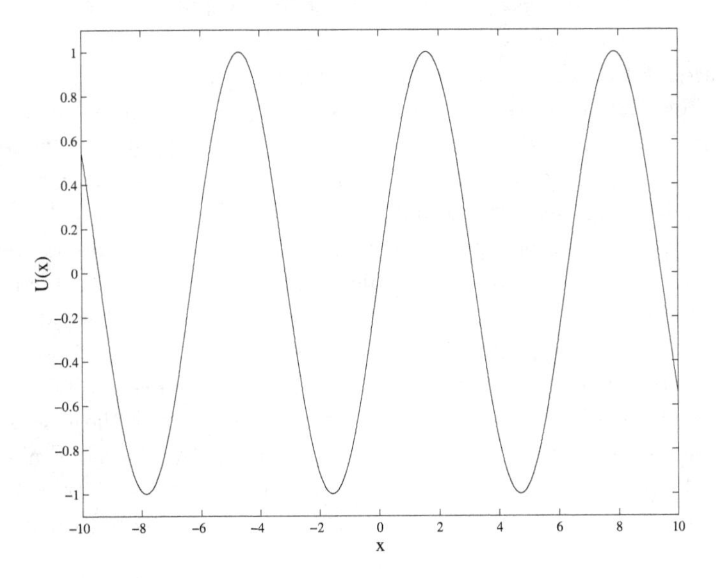

Fig. 4. Unity-bounded functon $\sin(x)$.

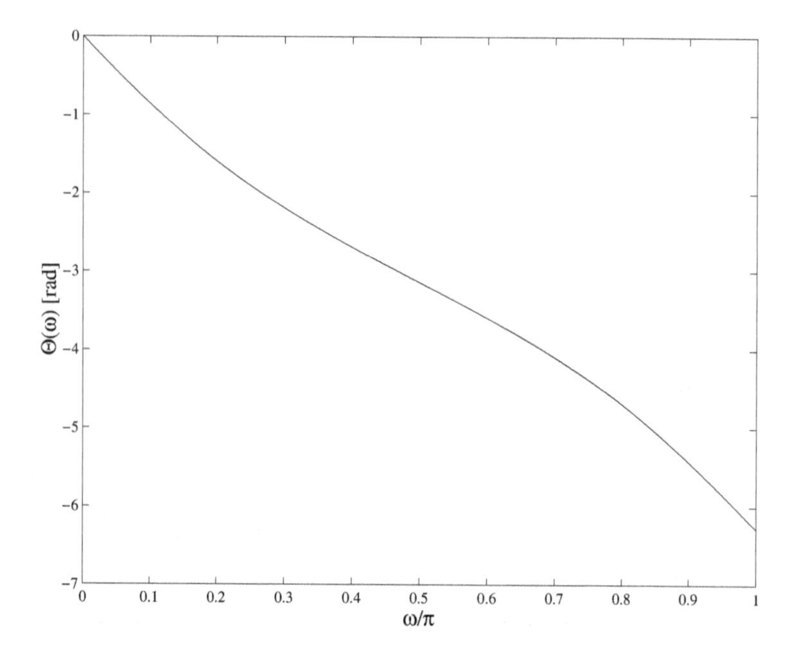

Fig. 5. Designed phase using the BL phase-error.

are used. Furthermore, the stability-margin requirement is $\rho = 0.9$, and the uniform sampled frequencies $\omega_1, \omega_2, \cdots, \omega_{1001}$ are employed.

The execution of the design program obtains the actual phase given in Fig. 5. To compare the ideal $\Theta_d(\omega)$ with the actual $\Theta(\omega)$, Fig. 6 shows the ideal $\Theta_d(\omega)$ (top) and the actual $\Theta(\omega)$ (bottom), and Fig. 7 shows both of the two in the same figure, where the deviations are indistinguishable (invisible). Fig. 8 shows the detailed phase deviations with peak value being 0.01223297134199, and Fig. 9 shows the detailed phase deviations in decibel (dB) with peak value being -38.2494 dB. The main goal of this BL-error-based design method is to guarantee that a preset stability margin (requirement) is truly met. Fig. 10 plots the locations of the two poles of the designed biquad system $H(z)$, and the radii of the two poles are given in Fig. 11. It is seen that the two poles have the same radius ($r = 0.4128$), which meets the preset requirement (the maximum value not greater than $\rho = 0.9$).

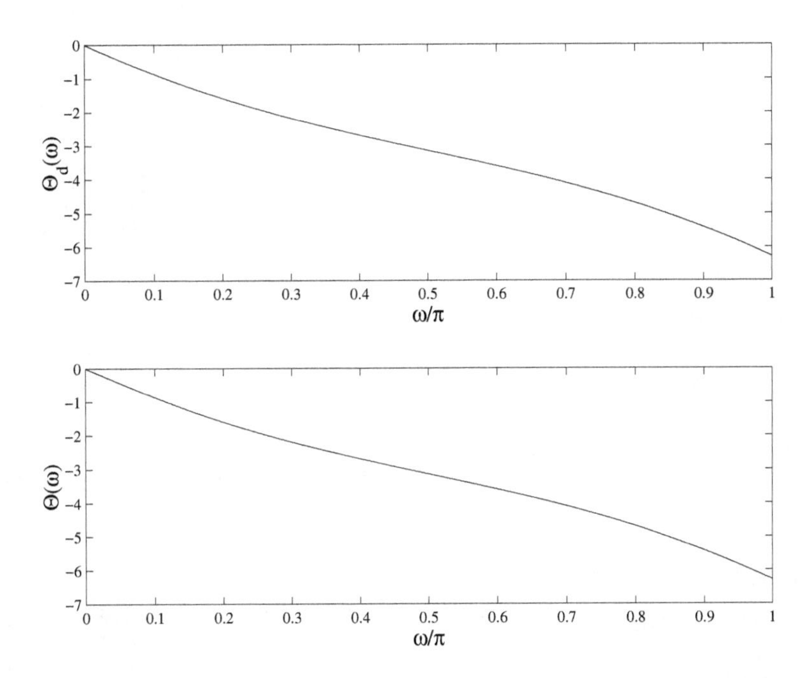

Fig. 6. Ideal phase (top) and actual phase (bottom).

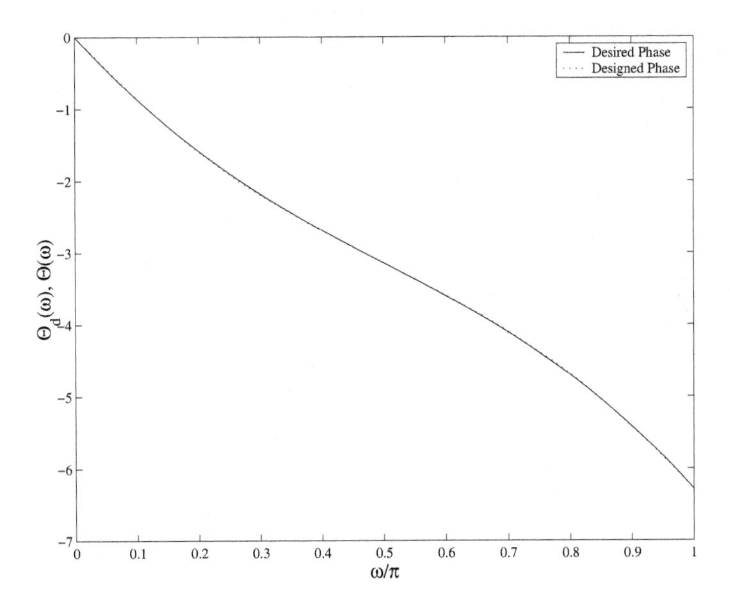

Fig. 7. Designed phase together with the ideal phase.

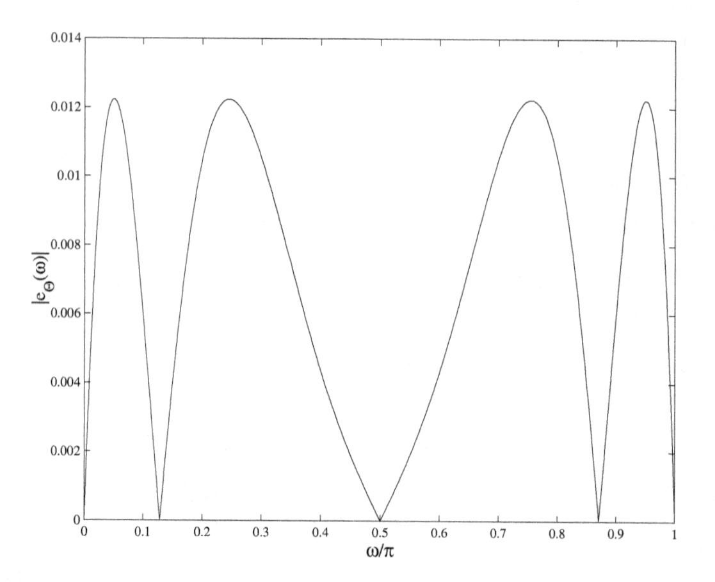

Fig. 8. Phase deviations.

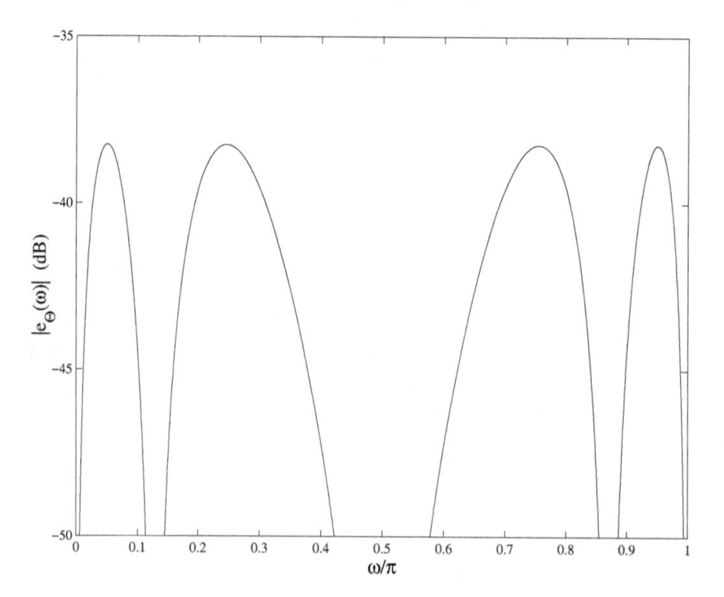

Fig. 9. Phase deviations in dB.

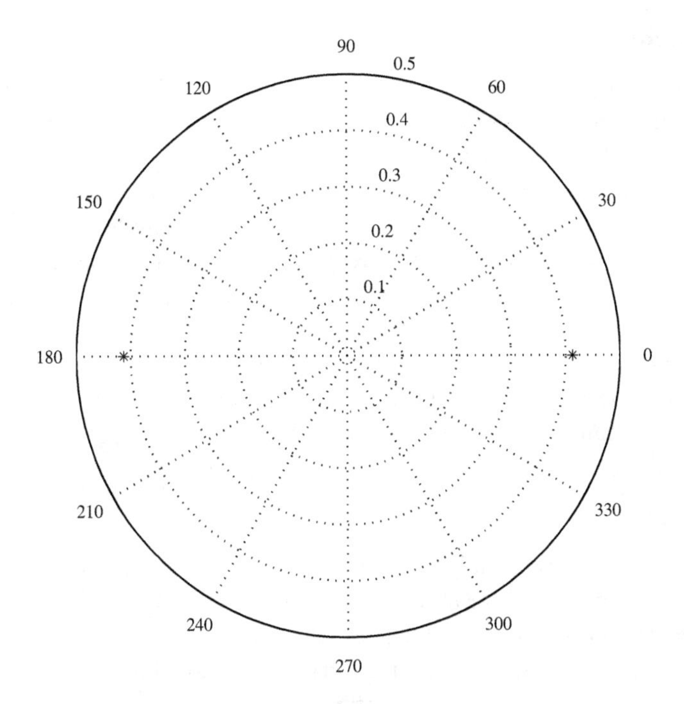

Fig. 10. Two poles within unit circle.

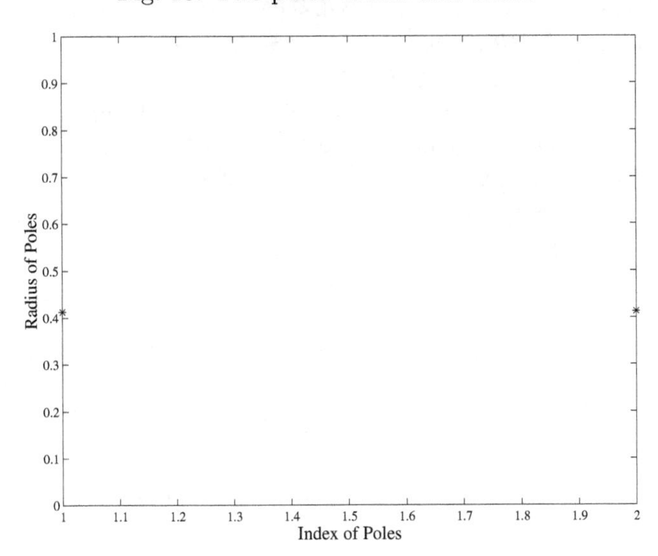

Fig. 11. Radii of the two poles.

5. Conclusion

This paper has applied the derived GST and BL phase-error formula to the biquad allpass system design. The designed biquad allpass phase-compensating system is useful for the phase compensation of a digital system with undesirable phase distortions. The GST has the favourable feature that enables the designer to design a biquad allpass system so as to satisfy a preset stability margin. The larger the margin is, the more robust the stability is. A large margin guarantees a strong robustness of the recursive system stability so that the system stability will not suffer from any coefficient-quantization errors, and the system is always stable. A biquad-system design example has illustrated the accuracy of the BL-error-based design algorithm as well as the achieved stability margin.

References

1. P. Soontornwong and S. Chivapreecha, "Pascal-interpolation-based noninteger delay filter and low-complexity realization," *Radioengineering*, vol. 24, no. 4, pp. 1002-1012, Dec. 2015.
2. T.-B. Deng, "Robust structure transformation for causal Lagrange-type variable fractional-delay filters," *IEEE Trans. Circuits Syst. I: Regular Papers*, vol. 56, no. 8, pp. 1681-1688, Aug. 2009.
3. T.-B. Deng, "Discretization-free design of variable fractional-delay FIR digital filters," *IEEE Trans. Circuits Syst. II, Analog Digit. Signal Process.*, vol. 48, no. 6, pp. 637-644, Jun. 2001.
4. T.-B. Deng, "Design of complex-coefficient variable digital filters using successive vector-array decomposition," *IEEE Trans. Circuits Syst. I, Reg. Papers*, vol. 52, no. 5, pp. 932-942, May 2005.
5. T.-B. Deng, "Minimax design of low-complexity even-order variable fractional-delay filters using second-order cone programming," *IEEE Trans. Circuits Syst. II: Express Briefs*, vol. 58, no. 10, pp. 692-696, Oct. 2011.
6. T.-B. Deng, S. Chivapreecha, and K. Dejhan, "Bi-minimax design of even-order variable fractional-delay FIR digital filters," *IEEE Trans. Circuits Syst. I: Regular Papers*, vol. 59, no. 8, pp. 1766-1774, Aug. 2012.
7. T.-B. Deng, "Noniterative WLS design of allpass variable fractional-delay digital filters," *IEEE Trans. Circuits Syst. I, Reg. Papers*, vol. 53, no. 2, pp. 358-371, Feb. 2006.
8. T.-B. Deng, "Generalized WLS method for designing allpass variable

fractional-delay digital filters," *IEEE Trans. Circuits Syst. I, Reg. Papers*, vol. 56, no. 10, pp. 2207-2220, Oct. 2009.

9. T.-B. Deng, "Minimax design of low-complexity allpass variable fractional-delay digital filters," *IEEE Trans. Circuits Syst. I, Reg. Papers*, vol. 57, no. 8, pp. 2075-2086, Aug. 2010.
10. T.-B. Deng, "Closed-form mixed design of high-accuracy all-pass variable fractional-delay digital filters," *IEEE Trans. Circuits Syst. I, Reg. Papers*, vol. 58, no. 5, pp. 1008-1019, May 2011.
11. N. Ito and T.-L. Deng, "Phase-compensation-circuit design using iterative linear-programming scheme," *Proc. IEEE ICCAS 2013*, pp. 5-8, Kuala Lumpur, Malaysia, Sept. 18-19, 2013.
12. T.-B. Deng, "Generalized stability-triangle for guaranteeing the stability-margin of the second-order digital filter," *Journal of Circuits, Systems, and Computers*, vol. 25, no. 8, 1650094 (13 pages), Aug. 2016.
13. T.-B. Deng, "Biquadratic digital phase-compensator design with stability-margin controllability," *Journal of Circuits, Systems, and Computers*, vol. 28, no. 4, 1950068 (15 pages), Apr. 2019.
14. T.-B. Deng, "Biquadratic phase-compensating system with robust stability," *Lecture Notes in Engineering and Computer Science: Proceedings of The International MultiConference of Engineers and Computer Scientists 2019, IMECS 2019*, Mar. 13-15, Hong Kong, pp. 176-179.

Road Usage Recovery in Fukushima Prefecture Following the 2011 Tohoku Earthquake

Jieling Wu
Iwate University, Morioka City, Iwate, Japan
j_wu@gw2k.hss.iwate-u.ac.jp

Noriaki Endo
Iwate University, Morioka City, Iwate, Japan
n_endo@gw2k.hss.iwate-u.ac.jp

We evaluate regional differences in road recovery in Fukushima Prefecture following the 2011 Tohoku Earthquake. We divided Fukushima Prefecture into seven regions, i.e., Soso, Iwaki, Kenhoku, Kenchu, Kennan, Aizu, and Minami-aizu regions. The cumulative usable road distance ratio of the main roads has been precisely calculated for each city from telematics probe-car data using the open source geographical information software. Defining the cumulative usable distance up to September 30, 2011 as 100%, the percentages of usable road distances were calculated. According to the results of our study, we conclude that the recovery conditions of regional roads in different areas of Fukushima Prefecture following the 2011 Tohoku Earthquake differed. The road recovery speeds in Coastal and Inland areas were not so different from each other. The road recovery speed in Aizu region was about a month slower than that in Coastal and Inland areas. The road recovery speed in Minami-aizu region was about a month slower than that in Aizu region. We concluded that the roads in these two regions whose recovery was significantly delayed were narrow, steep-walled, and located in mountainous regions.

Keywords: 2011 Tohoku Earthquake; big data analysis; ; Fukushima Prefecture; probe-car telematics data; vehicle-tracking map

1. Introduction

1.1. *The 2011 Tohoku Earthquake*

The 2011 Tohoku Earthquake [Fig. 1] struck the northeastern coast of Japan on March 11, 2011. Subsequently, the region was severely affected by the tsunami. Following these natural disasters, the electricity, water, and gas supplies were shut down in both coastal and inland areas.[1–5] Furthermore, the road travel was disrupted in many parts of the region. The tsunami inundation distance (the distance that a tsunami wave travels) and the run-up elevation (the highest elevation above sea level that tsunami waves reach)[6] were determined in a paper.[7]

1.2. *Purpose*

The primary purpose of our study was to evaluate the regional differences in road recovery in Fukushima Prefecture following the 2011 Tohoku Earthquake. Therefore, based on geographic position and features, we divided Fukushima Prefecture into seven regions, i.e., Soso, Iwaki, Kenhoku, Kenchu, Kennan, Aizu, and Minami-aizu regions.

During the disaster, these areas were affected differently. For example, due to differences in coastal features, the tsunami struck the northern coastal area more heavily than the southern coastal area. Therefore, we assumed that there were specific differences among the three studied regions during the road recovery process following the disaster.

The secondary purpose of our study was to compare the regional differences with regard to road recovery in Fukushima Prefecture, the target of this study, compared to road recovery in Iwate Prefecture and Miyagi Prefecture, which was evaluated in our previous studies.[3,4]

This paper is the revised and extended version of the previous paper[5] in the IMECS 2019 proceedings.

2. Our previous studies

2.1. *Previous study 1*

Our previous study 1[1] was focused on the use of the main roads in the Southern Coastal area of Iwate Prefecture. The usable distances of the main roads following the 2011 Tohoku Earthquake have been calculated from the G-BOOK telematics data.[8]

The main findings of this study are as follows:

1) The usable distance of the roads in a weekly period has continuously increased from March 18 to April 7, 2011, but it has fluctuated thereafter.

2) Defining the cumulative usable distance up to September 30, 2011 as 100%, it has been determined that 80% of the road distance has been usable by April 7, 2011 and 90% by April 29, 2011.

3) The use of the main road in the coastal area of Iwate Prefecture has been completely recovered by April 29, 2011.

2.2. *Previous study 2*

Our previous study 2[2,15] was focused on the use of the main roads not only in the Southern but also in the Northern Coastal area of Iwate Prefecture.

The cumulative usable road distance ratio of the main roads has been precisely calculated for each city using the free and open source geographical information system software QGIS. The main findings in this study[2] are listed below.

(1) The change in the cumulative usable road distance ratio during the research period differed from one city to the next.

(2) The ratio increases in the usable distances of Kuji, Iwaizumi, and Noda were extremely delayed.

In our study, we were able to determine related roads by analyzing the maps generated by the QGIS software. For Kuji and Iwaizumi, the road whose recovery was significantly delayed is the Iwate Prefectural Road number 7 (Kuji-Iwaizumi line). For Noda, the road whose recovery was significantly delayed is the Iwate Prefectural Road number 273 (Akka-Tamagawa line).

(3) In our previous study 1,[1] we determined that the use of the main road in the Southern Coastal area of Iwate Prefecture was completely recovered by April 29, 2011.

However, in this study, when we have precisely observed the change in the usable road distance ratio during the research period for each city, the ratio increase in the usable road distance of Kamaishi has been delayed compared with other Southern Coastal cities.

For Kamaishi City, the road whose recovery was significantly delayed is the Iwate Prefectural Road number 249 (Sakuratoge-Heita line).

2.3. *Previous study 3*

In our previous study,[3,16] we calculated the regional differences for road recovery in Iwate Prefecture following the 2011 Tohoku Earthquake. We divided Iwate Prefecture into four areas, i.e., Northern Inland, Southern Inland, Northern Coastal, and Southern Coastal areas. The main results of the previous study are as follows.

First, we determined that, for Northern and Southern Inland areas, 80% of the road distance was usable by April 15, 2011 and 90% by May 27, 2011, which indicates that the recovery speed in these areas was slightly slower than that in the Southern Coastal area.

Second, we found that, for the Northern Coastal area, 80% of the road distance was usable by April 29, 2011 and 90% by June 24, 2011, which implies that the recovery speed in the Northern Coastal area was significantly slower than that in the Southern Coastal area.

Hence, we concluded that these findings are related to the fact that road recovery efforts were more focused on regions heavily affected by the earthquake and tsunami.

2.4. *Previous study 4*

In our previous study,[4,17] we evaluate regional differences in road recovery in Miyagi Prefecture following the 2011 Tohoku Earthquake. We divided Miyagi Prefecture into three areas, i.e., Inland, Northern Coastal, and Southern Coastal areas. According to the results of our study, we conclude that the recovery conditions of regional roads in different areas of Miyagi Prefecture following the 2011 Tohoku Earthquake differed. In the Northern Coastal area, 80% of the road distance was usable by April 15, 2011 and 90% was usable by May 27, 2011. In the Southern Coastal area, 80% of the road distance was usable by March 31, 2011 and 90% was usable by April 8, 2011. Recovery in the Southern Coastal area was much faster compared to that in the Northern Coastal area. We assume that this is due to the shape of the coastlines. The coastlines in the Northern coastal area are primarily rias. The coastlines in the Southern coastal area are mostly sandy. Furthermore, we have concluded that the recovery conditions of the regional roads following the 2011 Tohoku Earthquake in the Northern Coastal area of Miyagi Prefecture were similar to those in the Southern Coastal area of Iwate Prefecture. In the disaster regions, similar recovery conditions were found according to geographic positions and features.

3. Telematics data and vehicle-tracking map

Telematics is a general term encompassing telecommunications and informatics. A telematics service provides various personalized information for users, especially for drivers of automobiles. G-BOOK is a telematics service provided by Toyota Motor Corporation.

To calculate the usable distance of the main roads, we applied the vehicle tracking map originally created by Hada et al.[9] after the 2007 Niigataken Chuetsu-oki earthquake.

That vehicle tracking map was based on telematics data provided by Honda Motor Company. Similarly, in our study, we used the vehicle tracking

Fig. 1. The epicenter of the 2011 Tohoku Earthquake occurred on March 11, 2011.

map based on telematics data provided by Toyota's G-BOOK system [Fig. 2].

Registered members of G-BOOK can access telematics services to acquire GPS data for car navigation systems and interactive driving data, such as traffic jam points, road closures, and weather reports.

Such comprehensive data acquisition is possible because the telematics system server receives accurate location data (geographic coordinates) from its registered members.

Telematics services are extremely useful to drivers. Because the accurate driving routes of registered users remain in the system server, they are accessible to traffic researchers in various fields.

4. Research Methods

4.1. *Research area*

The current study was focused on the entire area of Fukushima Prefecture (i.e., Soso, Iwaki, Kenhoku, Kenchu, Kennan, Aizu, and Minami-aizu regions) [Fig.3].

4.2. *Research materials*

In our current study, we have used the vehicle tracking maps built from the G-BOOK telematics data that is available on the Internet on March 18, 2011 following the 2011 Tohoku Earthquake.[8]

The data used in this study have been collected between March 18 and September 30, 2011 (i.e., approximately six months following the 2011 Tohoku Earthquake).

4.3. *System*

Hardware:

The computations have been performed on a standard PC laptop with a Core i7-6700U CPU (2.6 GHz) and 16 GB memory.

Software:

The software QGIS version 2.18.20 (the latest version available)[11] ,LibreOffice Calc 4.2.7 spreadsheet software,[12] and Microsoft Excel 2010 running on the Windows 7 Professional operating system have been used in this study. It is well-known that QGIS is one of the most popular geographic information systems used worldwide.

Prior to the abovementioned applications for geographical data processing, we have used the ogr2ogr software[13] on the Linux operating system along with Vine Linux 4.2,[14] which is a Linux distribution developed by a Japanese Linux community.

Note that QGIS, LibreOffice Calc, ogr2ogr, and Vine Linux are open source softwares freely available on the Internet.

4.4. *Data processing*

1) The vehicle tracking maps constructed from the G-BOOK telematics data have been provided in the Google map KMZ format. For our analysis, we have first converted the KMZ files to SHP files (i.e., shape-files), which are compatible with ArcGIS using the ogr2ogr software.

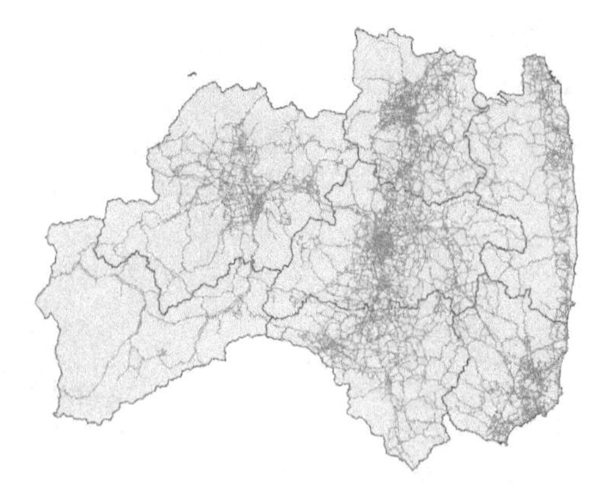

Fig. 2. Vehicle tracking map of Fukushima Prefecture

Fig. 3. Fukushima Prefecture divided into seven regions, i.e., Soso, Iwaki, Kenhoku, Kenchu, Kennan, Aizu, and Minami-aizu regions. The perimeter of a region is shown by a gray polygon.

2) Next, the data coordinates have been converted from the terrestrial latitude and longitude to the x and y coordinates in a rectangular coordinate system.

3) To reduce the computation time, the data file has been clipped to small files containing only the research area.

4) After merging daily data into weekly data and removing duplicate data, we have been able to calculate the exact usable road distance available for a given week.

In this context, a usable road is one on which at least one vehicle has been probed during the observation period.

The purpose of converting the daily data to weekly data was to smooth the daily fluctuations in the traffic flows.

5) Next, we have calculated the proportion of the cumulative distance up to the specified date. Note that the cumulative distance up to September 30, 2011 was considered 100%.

6) In this research, the date when this rate exceeded 90% was defined as the indicator of road usage recovery.

5. Results

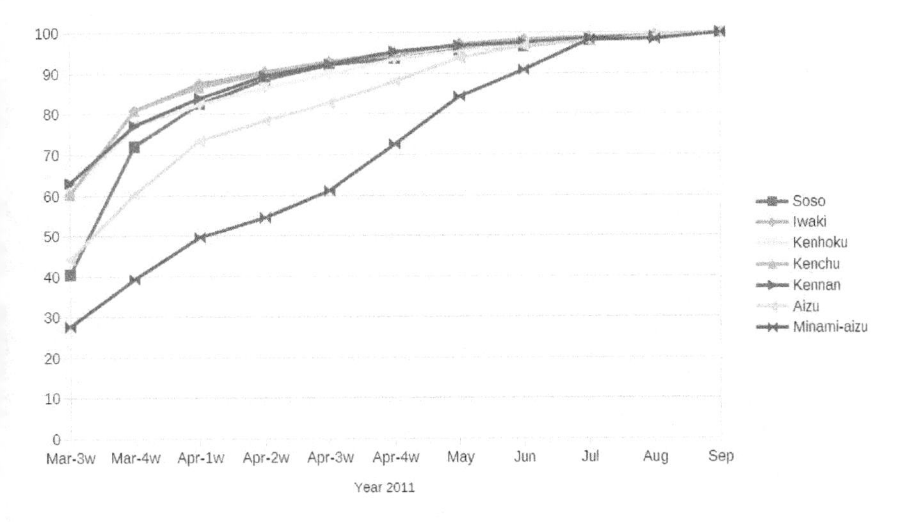

Fig. 4. Cumulative usable road distance ratio for each area. The vertical scale displays the cumulative distance proportion of the usable roads (relative to the cumulative distance on September 30, 2011) for each date.

Table I

Regional difference for road recovery in Fukushima Prefecture (cumulative usable road distances (kilometers) and ratios)

	Mar 24	Mar 31	Apr 08	Apr 15	Apr 22	Apr 29	May 27	Jun 24	Jul 29	Aug 26	Sep 30
Coastal (Hama-dori)	501.60	894.71	1023.22	1098.45	1143.27	1161.89	1188.78	1198.79	1217.53	1230.06	1240.69
Soso	40.4	72.1	82.5	88.5	92.1	93.6	95.8	96.6	98.1	99.1	100
Coastal (Hama-dori)	877.30	1176.65	1273.40	1316.11	1352.35	1375.66	1414.69	1429.99	1438.81	1444.68	1455.02
Iwaki	60.3	80.9	87.5	90.5	92.9	94.5	97.2	98.3	98.9	99.3	100
Inland (Naka-dori)	1150.32	1454.07	1555.05	1634.73	1696.24	1767.36	1804.52	1834.23	1855.28	1875.95	1889.07
Kenhoku	60.9	77.0	82.3	86.5	89.8	93.6	95.5	97.1	98.2	99.3	100
Inland (Naka-dori)	1549.99	2079.07	2229.02	2317.43	2390.14	2447.85	2494.98	2506.59	2532.57	2556.73	2571.73
Kenchu	60.3	80.8	86.7	90.1	92.9	95.2	97.0	97.5	98.5	99.4	100
Inland (Naka-dori)	906.70	1109.48	1206.45	1286.18	1327.17	1371.37	1391.99	1402.48	1416.81	1422.61	1437.81
Kennan	63.1	77.2	83.9	89.5	92.3	95.4	96.8	97.5	98.5	98.9	100
Far-inland	796.17	1084.43	1321.46	1410.88	1489.16	1585.62	1689.17	1743.03	1763.68	1784.55	1800.42
Aizu	44.2	60.2	73.4	78.4	82.7	88.1	93.8	96.8	98.0	99.1	100.0
Far-inland	130.29	184.69	233.22	256.31	287.31	340.83	396.12	426.97	461.83	462.66	469.88
Minami-aizu	27.7	39.3	49.6	54.5	61.1	72.5	84.3	90.9	98.3	98.5	100.0

Table II

Regional differences for road recovery in Miyagi Prefecture (cumulative usable road distances (meters) and ratios)[4]

	Mar 31	Apr 08	Apr 15	Apr 22	Apr 29	May 27	Jun 24	Jul 29	Aug 26	Sep 30
Whole	3710404.71	4070465.56	4302746.51	4432316.59	4555942.55	4689705.40	4784541.96	4853809.92	4898789.43	4951562.17
Inland	0.749	0.822	0.869	0.895	0.920	0.947	0.966	0.980	0.989	1.000
Northern	689499.47	760734.33	798356.91	820800.18	861329.64	884760.76	917857.84	938069.52	946266.16	959149.12
Coastal	0.719	0.793	0.832	0.856	0.898	0.922	0.957	0.978	0.987	1.000
Southern	1401296.10	1501843.72	1544385.42	1576106.64	1589184.41	1610740.67	1621919.88	1634768.72	1639012.45	1643175.95
Coastal	0.853	0.914	0.940	0.959	0.967	0.980	0.987	0.995	0.997	1.000

Table III

Regional difference for road recovery in Iwate Prefecture (cumulative usable road distances (meters) and ratios)[3]

	Mar 31	Apr 08	Apr 15	Apr 22	Apr 29	May 27	Jun 24	Jul 29	Aug 26	Sep 30
Northern	1850873.51	2070475.04	2152283.40	2251706.55	2296780.33	2403134.24	2483876.38	2532471.47	2571798.73	2603629.91
inland	0.711	0.795	*0.827*	0.865	0.882	*0.923*	0.954	0.973	0.988	1.000
Southern	2386739.37	2722218.13	2870724.95	2991566.34	3074037.86	3201569.09	3310978.41	3396994.03	3429255.89	3465835.57
inland	0.689	0.785	*0.828*	0.863	0.887	*0.924*	0.955	0.980	0.989	1.000
Northern	394082.91	449373.30	471389.15	524694.94	564884.21	589275.55	614317.97	618734.27	668603.07	675390.65
coastal	0.583	0.665	0.698	0.777	*0.836*	0.872	*0.910*	0.916	0.990	1.000
Southern	666655.53	779216.76	814157.35	821573.46	843427.24	855066.71	903310.83	910464.22	920120.72	934580.47
coastal	0.713	*0.834*	0.871	0.879	*0.902*	0.915	0.967	0.974	0.985	1.000

5.1. *Regional road recovery differences*

Defining the cumulative usable distance up to September 30, 2011 as 100%, the percentages of usable road distances are given in Table.II. In Table.II, the upper lines indicate the cumulative usable road distances (in kilometers), and the lower lines represent the ratio of cumulative usable road distance.

5.1.1. *Soso region of Coastal area (Hama-dori)*

It was determined that 80% of the road distance was usable by April 8, 2011 and 90% was usable by April 22, 2011.

5.1.2. *Iwaki region of Coastal area (Hama-dori)*

It was determined that 80% of the road distance was usable by March 31, 2011 and 90% was usable by April 15, 2011.

The recovery speed in Iwaki region was slightly faster than that in Soso region.

5.1.3. *Kenhoku region of Inland area (Naka-dori)*

It was determined that 80% of the road distance was usable by April 8, 2011 and 90% was usable by April 29, 2011.

The recovery speed in Kenhoku region was almost the same as that in Soso region (a very little slower).

5.1.4. *Kenchu region of Inland area (Naka-dori)*

It was determined that 80% of the road distance was usable by March 31, 2011 and 90% was usable by April 15, 2011.

The recovery speed in Kenchu region was almost the same as that in Iwaki region.

5.1.5. *Kennan region of Inland area (Naka-dori)*

It was determined that 80% of the road distance was usable by April 8, 2011 and 90% was usable by April 22, 2011.

The recovery speed in Kennan region was almost the same as that in Soso region.

As a whole, road recovery speeds of region one to five were not so different from each other.

5.1.6. *Aizu region of Far-inland area*

It was determined that 80% of the road distance was usable by April 22, 2011 and 90% was usable by May 27, 2011.

The recovery speed in Aizu region was about a month slower than that in Coastal and Inland area.

5.1.7. *Minami-aizu region of Far-inland area*

Minami means south in Japanese language. It was determined that 80% of the road distance was usable by May 27, 2011 and 90% was usable by June 24, 2011.

The recovery speed in Minami-aizu region was about a month slower than that in Aizu region.

6. Discussion

6.1. *Two far-inland regions whose road recovery was extremely delayed after the 2011 Tohoku Earthquake*

We could evaluate the road recovery speed after the 2011 Tohoku Earthquake clearly by observing Table 1 and Figure 4.

The recovery speed did not differ so much in the Coastal (Soso and Iwaki) and Inland (Kenhoku, Kenchu, and Kennan) regions. In these regions, each recovery difference was within a week or so.

But the recovery speed in Aize was significantly slower than that in the Coastal and Inland regions. And the recovery speed in Minami-aizu (located in the south of Aizu) was significantly slower than that in Aizu.

6.2. *Far-inland regions, namely Aizu and Minami-aizu*

We have watched some Youtube video images of the roads in these regions.

We concluded that the roads in these regions whose recovery was significantly delayed were narrow, steep-walled, and located in mountainous regions.

We would like to build the similar system which we had built in a previous study. The system would allow us to browse video images of such vulnerable roads easily.

6.3. *Comparison of regional differences for road recovery in Fukushima Prefecture, MiyagiPrefecture, and Iwate Prefecture*

According to the results of our study, we have concluded that the recovery conditions of the regional roads following the 2011 Tohoku Earthquake in the Aizu region of Fukushima Prefecture were similar to those in the Northern and Southern Inland area of Iwate Prefecture [Table 1 and 3].

Furthermore, the recovery conditions of the regional roads following the 2011 Tohoku Earthquake in the Iwaki region of Fukushima Prefecture were almost similar to those in the Southern Coastal area of Miyagi Prefecture [Table 1 and 2].

We had already discussed in our previous paper,[4] in the disaster regions, similar recovery conditions were found according to geographic positions and features. Now, it has become more clear in this research.

6.4. *A huge incident regarding the Fukushima Dai-ichi nuclear plant in Soso region*

In Soso region, there was a huge incident regarding the Fukushima Dai-ichi nuclear plant when the 2011 Tohoku Earthquake [Fig. 1] occurred.

Regarding this region, we need more precise analysis of traffic recovery, therefore we avoid further discussion of it in this research.

We would like to conduct that kind of analysis in the near future.

Acknowledgments

The authors are grateful to Ms. Junko Akita (Lecturer of Iwate University) for her contribution to our studies.

References

1. Hayato Komori and Noriaki Endo, "Road Distance Traveled by Vehicles Following the 2011 Tohoku Earthquake, Calculated by G-BOOK Telematics Data," *Proceedings of the 9th International Conference on Signal-Image Technology & Internet-Based Systems*, SITIS 9, Dec.2-5, 2013, Kyoto, Japan, 870-874.
2. Noriaki Endo and Hayato Komori, "Analysis of Vehicle Tracking Maps in Iwate Prefecture Following the 2011 Tohoku Earthquake," *Lecture Notes in Engineering and Computer Science: Proceedings of the International MultiConference of Engineers and Computer Scientists 2015*, IMECS 2015, Mar.18-20, 2015, Hong Kong, 124-128.

Available at http://www.iaeng.org/publication/IMECS2015/IMECS2015_pp124-128.pdf

3. Noriaki Endo and Hayato Komori, "Regional Difference on Road Recovery in Iwate Prefecture Following the 2011 Tohoku Earthquake," *Lecture Notes in Engineering and Computer Science: Proceedings of the World Congress on Engineering and Computer Science 2015*, WCECS 2015, Oct.21-23, 2015, Berkeley, CA, USA, 162-167.
   Available at http://www.iaeng.org/publication/WCECS2015/WCECS2015_pp162-167.pdf
4. Noriaki Endo, "Regional Differences in Miyagi Prefecture Road Recovery Following the 2011 Tohoku Earthquake," *Lecture Notes in Engineering and Computer Science: Proceedings of The International MultiConference of Engineers and Computer Scientists 2016*, IMECS 2016, Mar.16-18, 2016, Hong Kong, 151-156.
   Available at http://www.iaeng.org/publication/IMECS2016/IMECS2016_pp151-156.pdf
5. Jieling Wu, and Noriaki Endo, "Regional Differences in Fukushima Prefecture Road Recovery Following the 2011 Tohoku Earthquake," *Lecture Notes in Engineering and Computer Science: Proceedings of The International MultiConference of Engineers and Computer Scientists 2019*, IMECS 2019, 13-15 March, 2019, Hong Kong, pp164-169.
   http://www.iaeng.org/publication/IMECS2019/IMECS2019_pp164-169.pdf
6. Hawai'i Tsunami Education Curriculum: Kai E'e:
   http://discovertsunamis.org/tsunami_science_u6.html
7. Nobuhito Mori, Tomoyuki Takahashi, Tomohiro Yasuda, and Hideaki Yanagisawa, "Survey of 2011 Tohoku earthquake tsunami inundation and run-up," *Geophysical Research Letters*, Vol.38, No.7, April, 2011.
8. Vehicle Tracking Map Constructed by G-BOOK Telematics Data (In Japanese):
   http://g-book.com/pc/spot/Tohoku-Jishin.asp
9. Yasunori Hada, Takeyasu Suzuki, Hiroki Shimora, Kimiro Meguro, and Noriko Kodama: Issues and Future Prospect on Practical Use of Probe Vehicle Data for Disaster Reduction -Provision of the Vehicle Tracking Map in the 2007 Niigataken Chuetsu-oki Earthquake- (In Japanese), *Journal of Japan Association for Earthquake Engineering*, 9(2), 148-159, 2009.
   Available at http://www.jaee.gr.jp/stack/submit-j/v09n02/090211_paper.pdf
10. Vehicle Tracking Map Constructed by G-BOOK Telematics Data
    http://g-book.com/disasterMap/ALL/Tohoku-Jishin_YYYYMMDD.kmz
    (For example, Tohoku-Jishin_20110318.kmz. The data files of tracking map are still available on December 20, 2018.)
11. Official website of QGIS, A Free and Open Source Geographic Information System
    http://www.qgis.org/en/site/
12. Official Website of the LibreOffice Project

http://www.libreoffice.org/
13. Official website of GDAL: ogr2ogr
http://www.gdal.org/ogr2ogr.html
14. Official website of Vine Linux (In Japanese)
http://www.vinelinux.org/
15. Noriaki Endo and Hayato Komori, "Municipal Difference on Road Recovery in Iwate Prefecture Following the 2011 Tohoku Earthquake," IAENG Transactions on Engineering Sciences - Special Issue for the International Association of Engineers Conferences 2015, World Scientific, 2016, pp.226-235. [Revised and extended version of the paper[2]] https://doi.org/10.1142/9789813142725_0018
16. Noriaki Endo, and Hayato Komori, "Regional Differences in Iwate Prefecture Road Usage Recovery Following the 2011 Tohoku Earthquake," Transactions on Engineering Technologies - World Congress on Engineering and Computer Science 2015, Springer, 2017, pp.219-231. [Revised and extended version of the paper[3]] https://doi.org/10.1007/978-981-10-2717-8_16
17. Noriaki Endo, "Road Usage Recovery in Miyagi Prefecture Following the 2011 Tohoku Earthquake," IAENG Transactions on Engineering Sciences - Special Issue for the International Association of Engineers Conferences 2016, World Scientific, 2017, pp.111-125. [Revised and extended version of the paper[4]] https://doi.org/10.1142/9789813226203_0010

Sentiment Analysis on Twitter with Python's Natural Language Toolkit and VADER Sentiment Analyzer

Shihab Elbagir[†] and Jing Yang

College of Computer Science and Technology, Harbin Engineering University, Harbin, Heilongjiang 150001, China
[†]E-mail: shihabsaad@ yahoo.com

Twitter data sentiment analysis has gained considerable attention as a research topic. The ability to obtain information about a public opinion by analyzing Twitter data and automatically classifying their sentiment polarity has attracted researchers because of the concise language used in tweets. In this study, we aimed at classifying the sentiments expressed in Twitter data using the Valence Aware Dictionary for sEntiment Reasoner (VADER). However, since most earlier research focused on binary classification, we propose a multi-classification scheme to analyze tweets in this study. We used VADER to classify tweets related to the 2016 US election. The finding of the present study reveals that the VADER sentiment analyzer can perform a good results in detecting ternary and multiple classes and accurately classify people's opinion.

Keywords: Natural Language Toolkit (NLTK); Twitter; sentiment analysis; Valence Aware Dictionary and sEntiment Reasoner (VADER).

1. Introduction

Social media technologies exist in several different forms, such as blogs, business networks, photo sharing, forums, microblogs, enterprise social networks, video sharing networks, and social networks. As the number of social media technologies has increased, various online social networking services, such as Facebook, YouTube, and Twitter, have become popular because they allow people to express and share their thoughts and opinions about life events.

These networks enable users to have discussions with different people across the world and to post messages in the forms of texts, images, and videos [1], [2]. Moreover, social media are enormous sources of information for companies to monitor public opinion and receiving polls about the products they manufacture. Microblogging services have become the best known and most commonly used platforms. Furthermore, they have evolved to become significant sources of different types of information [3].

Twitter is a popular microblogging service that allows users to share, deliver, and interpret real-time, short, and simple messages called tweets [4]. Therefore, Twitter provides a rich source of data that are used in the fields of opinion mining and sentiment analysis. Recently, the sentiment analysis of Twitter data has attracted the attention of researchers in these fields. However, most state-of-the-art studies have used sentiment analysis to extract and classify information about the opinions expressed on Twitter concerning several topics, such as predictions, reviews, elections, and marketing.

Currently, many tools, such as Linguistic Inquiry and Word Count (LIWC) [5], offer the means of extracting advanced features from texts. However, most of these tools require some programming knowledge. In the present work, the Valence Aware Dictionary and sEntiment Reasoner (VADER) [6] is used to determine the polarity of tweets and to classify them according to multiclass sentiment analysis. The remainder of this paper is structured as follows: section 2 provides a brief description of related studies in the literature. In section 3, we present in detail the methodology. In section 4, we describe the tool used in this study. In section 5, we discuss the results. In section 6, we conclude and provide recommendations for future work [19].

2. Related Work

Recently, researchers have shown increasing interest in the field of sentiment analysis, particularly regarding Twitter data. The following are previous studies that have contributed to the field of sentiment analysis in the past few years. Wagh *et al.* [7] developed a general sentiment classification system for use if no label data are available in the target domain. In this system, labeled data in a different domain are used. Moreover, this system was used to calculate the frequency of each term in a tweet. In this study, a dataset containing four million tweets that were publicly available by Stanford University was analyzed. This dataset was used to predict the polarity of sentiments expressed in people's opinions. Traditional classification algorithms can be used to train sentiment classifiers from manually labeled text data, but the manually labeling work is expensive and time-consuming. The study found that if a classifier trained in one domain is applied directly to other domains, the performance is extremely low. The work showed the accuracy of different algorithms for different numbers of tweets, such as the following: Naive Bayes, Multi-nominal NB, Linear SVC, Bernoulli NB classifier, Logistic Regression, and the SGD classifier. The results showed that the proposed system was more efficient than the existing systems.

Gilbert [6] developed VADER, which is a simple rule-based model for general sentiment analysis and compared its effectiveness to 11 typical state-of-the-practice benchmarks, including Affective Norms for English Words(ANEW), Linguistic Inquiry and Word Count (LIWC), the General Inquirer, Senti WordNet, and machine learning-oriented techniques that rely on the Naive Bayes, Maximum Entropy, and Support Vector Machine (SVM) algorithms. The study described the development, validation, and evaluation of VADER. The researcher used a combination of qualitative and quantitative methods to produce and validate a sentiment lexicon that is used in the social media domain. VADER is utilizing a parsimonious rule-based model to assess the sentiment of tweets. The study showed that VADER improved the benefits of traditional sentiment lexicons, such as LIWC. VADER was differentiated from LIWC because it was more sensitive to sentiment expressions in social media contexts, and it generalized more favorably to other domains.

Mane *et al.* [8] presented a sentiment analysis using Hadoop, which quickly processes vast amounts of data on a Hadoop cluster in real time. The researchers aimed to determine whether the users expressed a positive or negative opinion. This approach was focused on the speed of performing sentiment analysis of real-time Twitter data using Hadoop. The Hadoop platform was designed to solve problems that involved large, unstructured, and complex data. It used the divide and rule method for processing such data. The overall accuracy of the project was determined by the time required to access various modules. In the analysis, the code yielded outstanding accuracy. The study used a numbering approach to rate the statements in multi-classes, which assigned a suitable range of different sentiments. Moreover, the approach could be used in other social media platforms, such as movie reviews (e.g., IMDB reviews) and personal blogs. Along the same line, Bouazizi and Ohtsuki [9] introduced SENTA, which helps users select from a wide variety of features those that are the best fit for the application used to run the classification. The researchers used SENTA to perform the multi-class sentiment analysis of texts collected from Twitter. The study was limited to seven different sentiment classes. The results showed that the proposed approach reached an accuracy as high as 60.2% in the multi-classification. This approach was shown to be sufficiently accurate in both binary classification and ternary classification.

Sanket *et al.* [10] proposed a novel approach to the evaluation of Twitter sentiment. They aim in this research to analyze Twitter data's lexicon features to classify it as positive, negative, or neutral. By implementing a spell checking algorithm to correct spelling mistakes in tweets, their technique enhances preprocessing. Furthermore, classifying tweets into their polarity of sentiment (positive, negative, or neutral). Based on the degrees of positivity or negativity, the researchers also group tweets [19].

3. Methodology

The current study consists of three phases. Phase one concerns the acquisition of Twitter data. Phase two focuses on the initial preprocessing work carried out to clean and remove irrelevant information from the tweets. Phase three deals with the use of the NLTK's VADER analyzer as well as the scoring method applied to the VADER results to assess its ability to classify tweets on a five-point scale. Figure 1. shows the architecture of our proposed sentiment classification system.

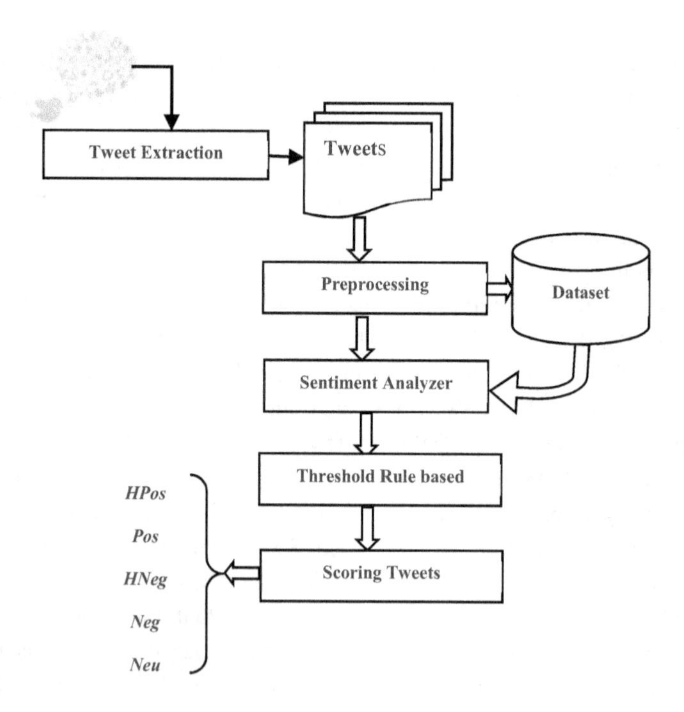

Fig. 1. The overall architecture of sentiment classifier.

3.1. *Data collection*

As aforementioned, the purpose of the data acquisition phase was to obtain Twitter data. The methods used to extract Twitter data allowed real-time access to publicly available raw tweets. To gather the data, we used Network Overview Discovery and Exploration for Excel (NodeXL) [11]. We collected a total of 2,430 political tweets concerning the 2016 US presidential election, which were published on Twitter's public message board and posted from 22 to 24 November 2016. Also, NodeXL, set the limit to a maximum of 2,000 tweets, from which we obtained a reduced data set. In order to collect the most relevant tweets, we used hashtags containing the candidates' names, Hillary and Trump. These names and "Election" were used as keywords to retrieve tweets, such as #Election Day results, #US Election 2016, #Election 2016, #Hillary Clinton, #Donald Trump. Table 1 illustrates the actual tweets retrieved by NodeXL [19].

Table 1. The raw tweets

	Tweet
1	NOT MY PRESIDENT | Anti Trump Tshirt | 24hr Shipping (small, red) https://t.co/RwmzsIh7us #Amazon #2016election #ElectionResults #ImWithHer
2	#WTFAmericaIn5Words #wtf #TrumpProtest #TrumpPresident #TrumpTrain #Election2016 #ElectionResults #election https://t.co/U8hM95xXR6
3	rt @michaelkokkinos: ???, ???... ?? ?????? ???????????, ?? ???? ! https://t.co/tyrxrm0cyc
4	RT @CharlesMBlow: Don't ask me to unify behind an unrepentant bigot. Can't do it. #ElectionResults
5	RT @paeznyc: It usually hits me the mornings. When does the sadness hit you? #ElectionResults https://t.co/5b6UkjJITU

3.2. *Data preprocessing*

A tweet is a microblog message posted on Twitter. It is limited to 140 characters. Most tweets contain text and embed URLs, pictures, usernames, and emoticons. They also contain misspellings. Hence, a series of preprocessing steps were carried out to remove irrelevant information from the tweets. The reason is that the cleaner the data, the more suitable they are for mining and feature extraction, which leads to the improved accuracy of the results. The tweets were also preprocessed to eliminate duplicate tweets and re-tweets from the dataset, which led to a final sample of 1,415 tweets. Each tweet was processed to extract its main message. To preprocess these data, we used Python's Natural Language Toolkit (NLTK). First, a regular expression (Regex) in Python was run to detect and discard tweets special characters, such as URLs ("http://url"), retweet (RT), user mention (@), and unwanted punctuation. Because hashtags (#) often explain the subject of the tweet and contain useful information related to the topic of the tweet, they are added as a part of the tweet, but the "#" symbol was removed.

Next, various functions of NLTK were used to convert the tweets to lowercase, remove stop words (i.e., words that do not express any meaning, such as is, a, the, he, them, etc.), tokenize the tweets into individual words or tokens, and stem the tweets using the Porter stemmer. When the preprocessing steps are complete, the dataset was ready for sentiment classification. Table 2 demonstrates the tweets after performing all the preprocessing steps [19].

Table 2. The preprocessing tweets

	Tweet
1	presid anti trump tshi 24hr ship small red amazon 2016elect election result im with h
2	wtfamericain5word wtf trump protest trump presid trump train election2016 election result
3	Tweet Removed
4	ask unifi behind unrepent bigot election result
5	usual hit morn sad hit election result

3.3. ***Sentiment analysis classification***

In this stage, the sentiments expressed in the tweets were classified. VADER Sentiment Analyzer was applied to the dataset. VADER is a rule-based sentiment analysis tool and a lexicon that is used to express sentiments in social media [6]. First, we created a sentiment intensity analyzer to categorize our dataset. Then the polarity scores method was used to determine the sentiment. The VADER Sentiment Analyzer was used to classify the preprocessed tweets as positive, negative, neutral, or compound. The compound value is a useful metric for measuring the sentiment in a given tweet.

3.3.1. *Threshold rule-based*

As mentioned above, the compound value used to categorize tweets as either positive, negative, or neutral. We consider a compound threshold value to be between 0.001 and −0.001. In the proposed method, a tweet with a compound value greater than the threshold was considered a positive tweet, and a tweet with a compound value less than the threshold was considered a negative tweet. In the remaining cases, the tweet was considered neutral. For classifying tweets as either positive, neutral, or negative. Algorithm 1 provides typical threshold values as:

Algorithm 1 Threshold Rule-based

Begin

Input: a tweet compound value

Output: a tweet with a score value

Function Threshold Values (compound):

For each tweet in pre-processing tweets **Do**:

IF compound value > Threshold **Then:**

assign score = 1 ⟹ (Positive)

IF (compound value > - Threshold) and (compound score < Threshold) **Then:**

assign score = 0 ⟹ (Neutral)

IF compound value < - Threshold **Then**:

assign score = -1 ⟹ (Negative)

End IF

End Function

End

3.3.2. *Tweet scoring rule*

Consequently, we apply a scoring rule classifier to classify the overall sentiment polarity in each tweet into high positive, positive, neutral, negative, and high negative based on the score value and positive or negative values. In the proposed method, the scoring rule is used to classify tweets into five sentiment classes as follows:

Algorithm 2 Tweet Scoring Rule

Begin

Input: a tweet score, positive, negative value

Output: overall sentiment score of the tweet

Function Tweet Sentiment Polarity (polarity):

For each tweet in pre-processing tweets **Do**:

IF (score value) = 1 **Then**:
If (positive value > 0.5) **Then:**

```
                tweet polarity = +2
            Else positive value < 0.5:
                tweet polarity = +1
              End IF
          End IF
          IF (score value) = -1 Then:
            If (negative value > 0.5) Then:
                tweet polarity = -2
            Else negative value < 0.5:
              tweet polarity = -1
            End IF
          End IF
          IF (score value) = 0 Then:
            tweet polarity = 0
          End IF
        End Function
      End
```

The polarity value gives the overall sentiment polarity of the tweet. The polarity value contains highly positive, positive, neutral, negative, and highly negative between +2 (highly positive) and -2 (highly negative). Polarity value is remapped to five coarse numerical categories: +2 (highly positive), +1 (positive), 0 (neutral), -1 (negative), and -2 (highly negative). We define this remapping based on intervals as in Eq. (1).

$$\text{Tweet polarity (value)} = \begin{cases} +2 & \textit{if} \text{ value is classified as } \textit{Highly Positive} \\ +1 & \textit{if} \text{ value is classified as } \textit{Positive} \\ -2 & \textit{if} \text{ value is classified as } \textit{Highly Negative} \\ -1 & \textit{if} \text{ value is classified as } \textit{Negative} \\ 0 & \textit{if} \text{ value is classified as } \textit{Neutral} \end{cases} \tag{1}$$

4. Tools

4.1. *Python*

Python is a programming language of interpretation, high-level, general-purpose. Python has a design philosophy that emphasizes code readability, particularly by using significant white space. It offers constructs that enable clear programming on both small and large scale. Python has a type dynamic system and an automatic memory management scheme. It supports various programming paradigms, including object-oriented, imperative, functional, and fully supports structured programming, and has an extensive standard library [15], [16].

4.2. *NodeXL*

NodeXL Basic is a free and open-source analysis and visualization software package that is used with Microsoft Excel [11]. This popular package is similar to other network visualization tools, such as Pajek, UCINET, and Gephi. NodeXL allows the quick and accessible collection of social media data through a set of import tools that gather data from social networks, such as Flickr, YouTube, Facebook, and Twitter. NodeXL focuses on the collection of publicly available data, such as Twitter statuses, and it follows the relationships of users who have made their accounts public. These features allow NodeXL users to instantly retrieve relevant social media data and integrate aspects of these data and their analysis into one tool [12].

4.3. *Natural Language Toolkit (NLTK)*

NLTK is a free open-source Python package that provides several tools for building programs and classifying data. NLTK is suitable for linguists, engineers, students, educators, researchers, and developers who work with textual data in natural language processing and text analytics [13]. NLTK provides an easy way to use the interfaces of over 50 corpora and lexical resources. It includes a group of text processing libraries for classification, tokenization, stemming, tagging, parsing, and semantic reasoning [14].

4.4. *Valence Aware Dictionary and sEntiment Reasoner (VADER)*

VADER is a lexicon- and rule-based sentiment analysis tool designed specifically to the sentiments expressed in social media and micro-blog data and achieving efficient and generalizable results in comparison with other state-of-the-art techniques. It is an open source tool that is completely free. VADER also requires word order and degree modifiers into consideration. VADER Providing polarity and intensity as well as validation by human experts. It also provides additional popular dictionary terms on emoticons, slang, and acronyms [6].

5. Results and Discussion

This section discusses the results of a Twitter sentiment analysis using NLTK and VADER sentiment analysis tools and the overview of the results of the classification. We use the built-in NLTK VADER Sentiment Analyzer to produce a Sentiment Intensity Analyzer (SIA) to categorize our tweets, then use the polarity scores method to get the sentiment. Figure 2 Shows each tweet's sentiment score as positive, negative, neutral, or compound as the VADER Sentiment Analyzer acquired.

```
[{'compound': -0.1531,
  'neg': 0.164,
  'neu': 0.714,
  'pos': 0.121,
  'tweet': 'presid anti trump tshi 24hr ship small red amazon 2016elect election
result im with'},
 {'compound': -0.5859,
  'neg': 0.352,
  'neu': 0.648,
  'pos': 0.0,
  'tweet': 'wtf americain 5 word wtf  trump protest trump presid trump train
election2016 election result 'elect'},
 {'compound': 0.0,
  'neg': 0.0,
  'neu': 1.0,
  'pos': 0.0,
  'tweet': 'ask unifi behind un repent bigot election result'}]
```

Fig. 2. Sentiment score of tweets using the VADER.

Figure 2 result consists of four values from the sentiment scoring such as Neutral, Negative, Positive, and Compound. The first three are the ratio of each category's sentiment score in our tweet, and the compound value that scores the sentiment. The compound score is normalized to be between -1 (Extremely Negative) and +1 (Extremely Positive).

Table 3. The tweets classification

	compound	neg	neu	pos	tweet	score
0	-0.1531	0.164	0.714	0.121	presid anti trump tshi 24hr ship small red ama...	-1
1	-0.5859	0.352	0.648	0.000	wtfamericain5word wtf trumpprotest trumppresid...	-1
2	0.0000	0.000	1.000	0.000	ask unifi behind unrepent bigot electionresult	0
3	0.3182	0.000	0.813	0.187	tom obama pardon clinton way presidenttrump el...	1
4	0.3612	0.000	0.783	0.217	realiz electionresult feel like rick trump neg...	1

The previous table demonstrates that the tweet classification is either positive, neutral or negative after setting the threshold rule for classification of tweets. In the current study, a compound threshold above 0.001 will be considered as positive and less than -0.001 as negative and, in another case, neutral. Results from Table 3 indicate that if a suitable threshold value is selected, we can achieve significantly compact results, and observation is to be made here, the size of the test data set will reduce considerably if we select a greater threshold value. Figure 3 shows the ternary percentages of polarity-based sentiment classification for each tweet as either positive, negative, or neutral.

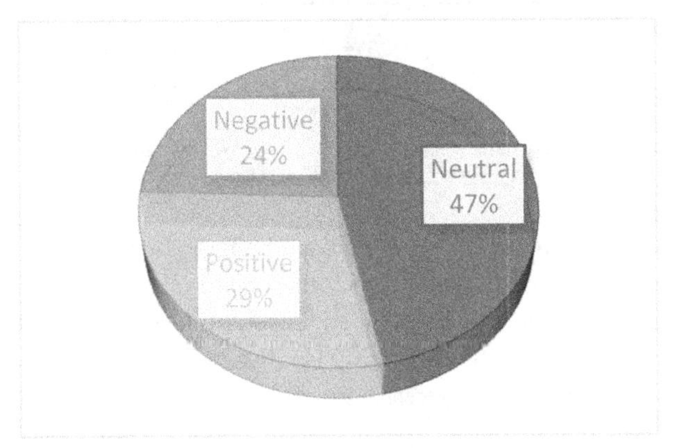

Fig. 3. The ternary classification

Table 4 shows the overall sentiment score and polarity of each tweet, taking into consideration the result of applying the tweet scoring rule.

Table 4. Overall sentiment polarity for every tweet

Tweets	label	Polarity
first, ignore laugh fight win gandhi everi true	+2	Highly Positive
When ev sad sorrow come watch conanobrien elect...	-2	Highly Negative
Unit state hay election 2016 election result ima...	0	Neutral
tom obama pardon Clinton way president trump el...	+1	Positive
play dead soon cat election result	-1	Negative

The result in Table 5 summarizes the distribution of tweets and the percentages of each of the five classes of sentiment. Most of the tweets in our dataset expressed negative or neutral opinions about the presidential election based on the results shown in Table 5. Interestingly, 26.5% of the tweets expressed positive opinions, and 22.89% of the tweets expressed negative opinions. A notable unbalance can be observed between the number of highly positive, positive, neutral, negative, and highly negative tweets. The neutral tweets are noticeably the highest (i.e., 46.7 %), among all other classes due to the small volume of tweets, which led to unbalanced data, and the assumption that the threshold value could provide a large number of neutral opinions. Also may have been based on the use of a general lexicon to categorize the political data.

Table 5. Polarity count and percentage for each class

Polarity	Count	Percentage
Highly Positive	34	2.402827
positive	375	26.501767
neutral	661	46.713781
negative	324	22.897527
Highly negative	21	1.484099

According to a count of tweets in each class, Figure 4 graphically represents the sentiment analysis polarity about people's opinion towards the 2016 US presidential election. Figure 4 shows the tweet counts in each class of sentiments. Highly negative, highly positive, negative, positive, and neutral obtained count of tweets 21, 34, 320, 366, and 673, respectively.

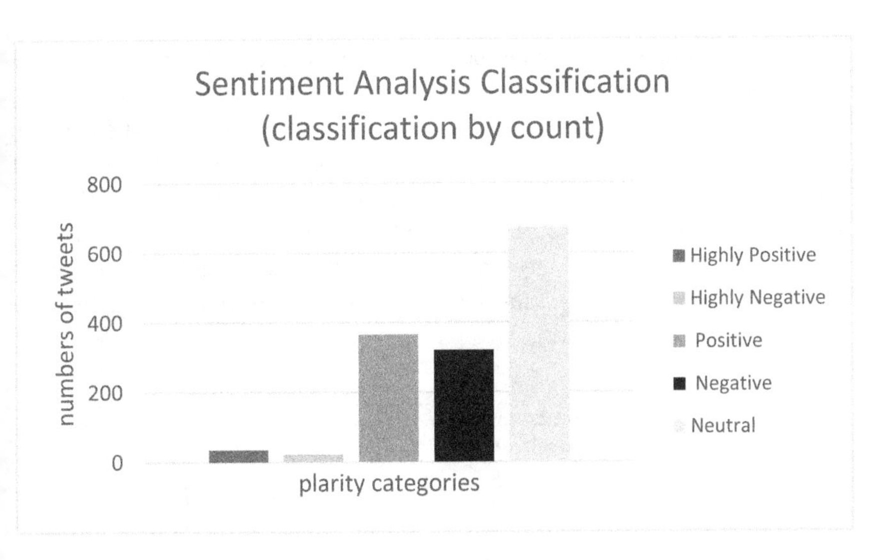

Fig. 4. Sentiment analysis classification by count

Figure 5 shows the percentage of the classification of sentiment analysis based on polarity for each class. The figure shows that the percentage obtained for Highly Positive is 2.40 %, but the highly negative percentage is 1.48 %. The percentages of positive tweets and negative tweets are 26.5%, 22.89% respectively. The percentage of neutral tweets is the highest, reaching 46.71 %.

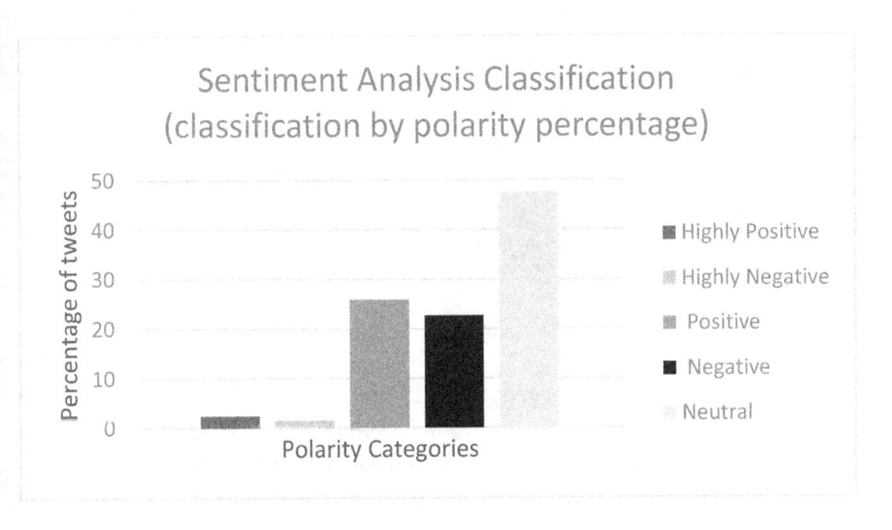

Fig. 5. Sentiment analysis based on polarity percentage

The most common positive words used in the dataset are shown in Figure 6. We can observe that most of the top words deal with the election result, Trump, election 2016, and elect. Figure 7 shows a graphical representation of the distribution of frequencies and attempts to examine the pattern of words and not each word specifically. The figure shows that our data fits within the Zipf's Law. The law of Zipf is an empirical law based on mathematical statistics. The law states that the frequency of any word is inversely proportional to its rank in the frequency table given a large sample of words used. The word number n therefore has a frequency of 1/n [17,18].

Positive word distributions:
[('election result', 381), ('election2016', 97), ('trump', 73), ('elect', 45), ('vote', 44), ('like', 33), ('amp', 33), ('popular', 26), ('win', 24), ('support', 21)]

Fig. 6. Most Common positive Words

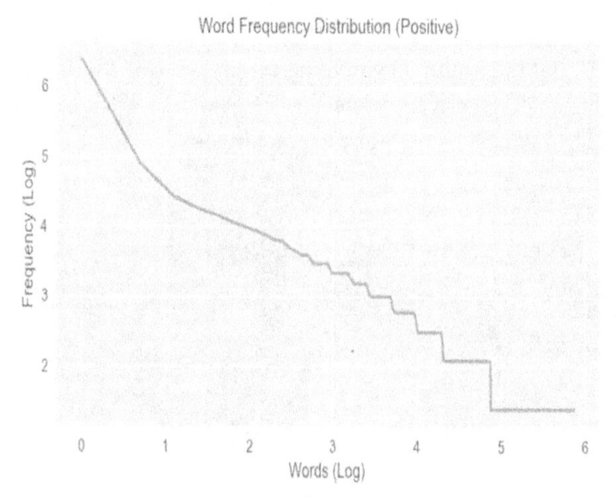

Fig. 7. Frequency Distributions of Positive

Figure 8 displays the most negative word frequency used in the dataset. Seeing that the "election results", "Trump", and "protest" are some of the top negative words. Interestingly, in the most common positive and negative words, we can see some of the same words, such as "election results", "election 2016" and "Trump".

Negative word distributions:
[('election result', 321),
('trump', 76),
('election2016', 63),
('protest', 48),
('elect', 41),
('amp', 32),
('people', 28),
('vote', 26),
('hillari', 21),
('riot', 21)]

Fig. 8. Most Common negative Words

Negative distribution also fits under the Zipf Law. In other words, Figure 9 shows that a vast minority of words appear most in our word distribution, whereas most words appear less.

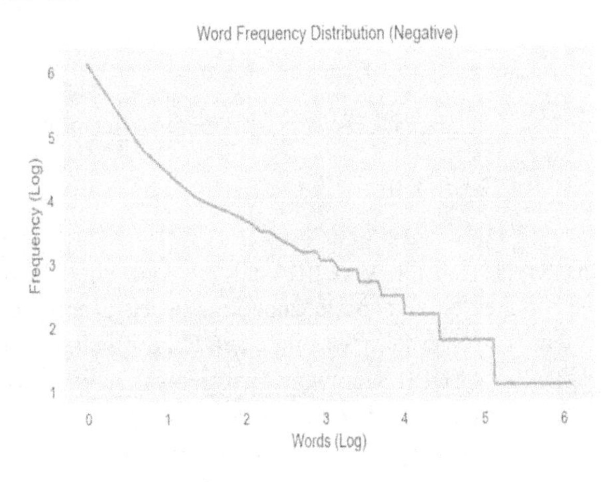

Fig. 9. Frequency Distributions of Negative

We can also grab from our dataset all the neutral words to get the most common neutral words as shown in Figure 10. While Figure 11 shows how frequently neutral words are distributed. The figure shows that our data fits under the Zipf Law. Looking at the result in the figures indicating that in our word distribution a vast minority of words appears most, whereas most words appear less.

Neutral word distributions:
[('election result', 633),
('election2016', 178),
('trump', 127),
('elect', 63),
('vote', 52),
('clinton', 28),
('amp', 28),
('not my presid', 26),
('trump protest', 24),
('us', 22)]

Fig. 10. Most Common neutral Words

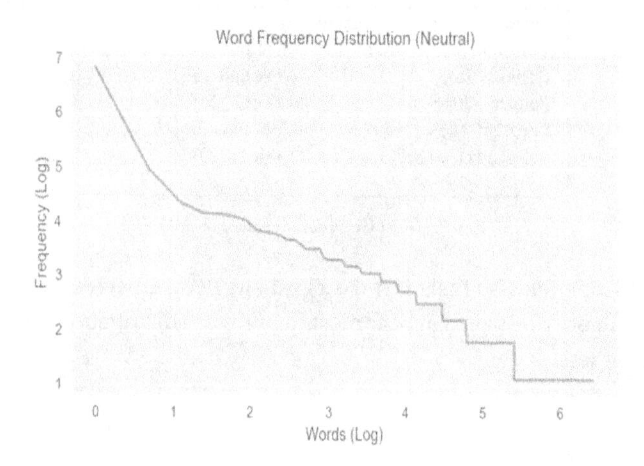

Fig. 11. Frequency Distributions of Neutral

6. Conclusion

In this study, the NLTK and the VADER analyzer were applied to conduct a sentiment analysis of Twitter data and to categorize tweets according to a multi-classification system. The case study was the 2016 US presidential election. The results indicated that the VADER Sentiment Analyzer was an effective choice for sentiment analysis classification using Twitter data. VADER easily and quickly classified huge amounts of data. However, the present study has the following limitations. First, a small volume of data was used. Second, a general lexicon was used to categorize specific data. Third, the data were not trained. In future work, we will improve our system by using large volumes of data, a specific lexicon, and a corpus for training the data to obtain good results.

Acknowledgments

This work was supported by National Natural Science Foundation of China (Nos. 61672179, 61370083, 61402126), Heilongjiang Province Natural Science Foundation of China (No. F2015030), Science Fund for Youths in Heilongjiang Province (No. QC2016083) and Postdoctoral Fellowship in Heilongjiang Province (No. LBHZ14071).

References

1. B. J. Jansen, M. Zhang, K. Sobel, and A. Chowdury, "Twitter power: Tweets as electronic word of mouth," J. Am. Soc. Inf. Sci. Technol., vol. 60, no. 11, pp. 2169–2188, 2009.
2. V. Kharde and P. Sonawane, "Sentiment analysis of twitter data: a survey of techniques," arXiv Prepr. arXiv1601.06971, 2016.
3. P. Selvaperumal and D. A. Suruliandi, "a Short Message Classification Algorithm for Tweet Classification," Int. Conf. Recent Trends Inf. Technol., pp. 1–3, 2014.
4. T. Singh and M. Kumari, "Role of Text Pre-processing in Twitter Sentiment Analysis," Procedia Comput. Sci., vol. 89, pp. 549–554, 2016.
5. Y. R. Tausczik and J. W. Pennebaker, "The psychological meaning of words: LIWC and computerized text analysis methods," J. Lang. Soc. Psychol., vol. 29, no. 1, pp. 24–54, 2010.
6. C. J. H. E. Gilbert, "Vader: A parsimonious rule-based model for sentiment analysis of social media text," in Eighth International Conference on Weblogs and Social Media (ICWSM-14). Available at (20/04/16) http://comp.social.gatech.edu/papers/icwsm14.vader.hutto.pdf, 2014.
7. B. Wagh, J. V Shinde, and P. A. Kale, "A Twitter Sentiment Analysis Using NLTK and Machine Learning Techniques," Int. J. Emerg. Res. Manag. Technol., vol. 6, no. 12, pp. 37–44, 2018.
8. S. B. Mane, Y. Sawant, S. Kazi, and V. Shinde, "Real Time Sentiment Analysis of Twitter Data Using Hadoop," Int. J. Comput. Sci. Inf. Technol., vol. 5, no. 3, pp. 3098–3100, 2014.
9. M. Bouazizi and T. Ohtsuki, "A Pattern-Based Approach for Multi-Class Sentiment Analysis in Twitter," IEEE Access, vol. 3536, no. c, pp. 1–21, 2017.
10. S. Sahu, S. K. Rout, and D. Mohanty, "Twitter Sentiment Analysis--A More Enhanced Way of Classification and Scoring," in 2015 IEEE International Symposium on Nanoelectronic and Information Systems (iNIS), 2015, pp. 67–72.
11. M. A. Smith et al., "Analyzing (social media) networks with NodeXL," in Proceedings of the fourth international conference on Communities and technologies, 2009, pp. 255–264.
12. D. L. Hansen, B. Shneiderman, and M. A. Smith, Analyzing social media networks with NodeXL: Insights from a connected world. Morgan Kaufmann, 2010.

13. S. Bird, E. Klein, and E. Loper, Natural language processing with Python: analyzing text with the natural language toolkit. "O'Reilly Media, Inc.," 2009.
14. http://www.nltk.org/ (Date Last Accessed, November 20, 2018).
15. D. Kuhlman, A python book: Beginning python, advanced python, and python exercises. Dave Kuhlman Lutz, 2009.
16. T. Peters, "Pep 20–the zen of python," Python Enhanc. Propos., 2004.
17. George, K. 1935. "Zipf. The Psychobiology of Language."
18. Zipf, George Kingsley. 1949. "Human Behavior and the Principle of Least Effort."
19. *Shihab Elbagir, and Jing Yang, "Twitter Sentiment Analysis Using Natural Language Toolkit and VADER Sentiment," Lecture Notes in Engineering and Computer Science: Proceedings of The International MultiConference of Engineers and Computer Scientists 2019, 13-15 March, 2019, Hong Kong, pp12-16.*

Analyzing the Tendency of Review Expressions for the Automatic Construction of Cosmetic Evaluation Dictionaries*

Mayumi Ueda[1], Yuna Taniguchi[2], Yuki Matsunami[3],
Asami Okuda[3] and Shinsuke Nakajima[4]

[1]*Faculty of Economics, University of Marketing and Distribution Sciences, Kobe, Hyogo, 651-2188, Japan. E-mail: Mayumi_Ueda@red.umds.ac.jp*

[2]*Faculty of Computer Science and Engineering, Kyoto Sangyo University, Kyoto, 603-8555, Japan. E-mail:g1644763@cc.kyoto-su.ac.jp*

[3]*Graduate School of Frontier Informatics, Kyoto Sangyo University, Kyoto, 603-8555, Japan. E-mail:i1888033@cc.kyoto-su.ac.jp*

[4]*Faculty of Information Science and Engineering, Kyoto Sangyo University, Kyoto, 603-8555, Japan. E-mail:nakajima@cc.kyoto-su.ac.jp*

Customer Reviews play an important role in the online shopping experience, particularly for products such as Cosmetics where the value of the product tends to be subjective. However, it is time consuming for consumers to filter through the large number of reviews and identify those which are meaningful to their purchase decisions. Recommender systems could be a useful tool in helping recommend products to customers which are relevant to their particular needs and context. However, there are many different attributes of the cosmetic product (such as anti-ageing effect and whitening effect) which could not be represented with a single user score. As such, In our research, we aim to develop a recommender system which takes into account such attributes when recommending cosmetic products to consumers. This paper describes how a cosmetic evaluation dictionary could be constructed based on review expressions extracted from text-based review comments for different cosmetic categories.

Keywords: cosmetic review, review analysis, automatic review rating, evaluation expression dictionary.

*This work was supported by the MEXT Grant-in-Aid for Scientific Research(C) (#16K00425, #19K12243) and Research(B)(#17H01822).

1. Introduction

There are many portal sites to support online shopping. These sites provide many kinds of information, such as item information by the company, cost, user-provided review, ranking information, and so on. Consumers can get effective information by user-provided reviews before they purchase them. In particular, consumers make careful choice about cosmetics since unsuitable items frequently cause skin irritations, therefore, they try to get effective information for themselves, through the online reviews.

@cosme[1] is a portal site that is very popular among Japanese women that provides product information and user reviews. While this site can be helpful for decision making, it is not easy to find truly suitable cosmetics. This is because, most of the existing review sites provide user reviews that contain review text and overall score and users have to read through huge volumes of review texts to find truly effective information for themselves. In this situation, it is difficult to understand the whole context of target cosmetic items intuitively. Therefore, we design and evaluate a collaborative recommender system for cosmetic items, which incorporates opinions of similar-minded users and automatically scores fine grained aspects of product reviews[2,3].

In order to develop such a review recommender system, we have to analyze review text to understand feedback on reviewers' experiences of cosmetic items. While there is a score of each review text on the conventional cosmetic review sites, It mostly consists of overall score, so this makes it difficult to recognize feedback on reviewers' experiences of cosmetic items from it. For example, there are "moisturizing effect" "whitening effect", "exfoliation care effect", "hypoallergenic effect", "aging care effect", etc. as aspects of reviews for "face lotion". Thus, we need a scoring method of such various aspects on cosmetic item review texts to understanding feedback on reviewers' experience of them.

Hence, the purpose of our study is to develop a method for automatic scoring of various aspects of cosmetic item review texts based on evaluation expression dictionary. The method can realize an automatic scoring of various aspects of cosmetic item reviews even if there are no scores for the individual aspects. As a first step, we constructed an evaluation expression dictionary for "face lotion", which has different feedback with each person manually[2]. According to simple experiments carried out, we can get the results of automatic scoring by our method which are quite close to the results of manual scoring from ground truth data. However, there are

Table 1. Categories of Cosmetic Item.

Skin care	Face wash, Makeup remover & Cleansers, Lotion, Emulsion, Serum, Face masks & Packs, Eye & Lip care
Makeup	Brow products, Eyeliner, Mascara, Eye shadow, Lips, Blush, Nails
Base makeup	Foundation, Base foundation, Concealer, Face powder

many categories for cosmetic items, for example, "face lotion", "emulsion", and "serum" as skin care items (see TABLE 1). It takes a lot of man-hours to develop evaluation expression dictionary for each categories of cosmetic items manually. Therefore, in this paper, we analyze the tendency of evaluation expression for each categories of cosmetic items for the purpose of developing the evaluation expression dictionaries automatically.

This paper is a revised version of the conference paper that we presented at IMECS2019[4].

The paper is organized as follows. The related work is given in Section 2. Then Section 3 describes the evaluation expression dictionary for automatic scoring of various aspects of cosmetic items. We analyze the features of evaluation expression for each cosmetic categories in Section 4. We conclude the paper in Section 5.

2. Related Work

There are several websites that provide reviews by consumers. For example, Amazon.com[5] and Priceprice.com[6] are popular shopping sites on the Internet which provide customer reviews of their merchandise. Moreover, "Tabelog" is a popular website in Japan. This website provides information and reviews of restaurants. In addition to the algorithmic aspects, researchers have recently focused on the presentation aspects of the review data[7].

Furthermore, in recent years, "@cosme" has become a very popular portal site for beauty and cosmetic items, and it provides a variety of information, such as reviews and shopping information regarding cosmetic items. According to the report issued by the company that operates "@cosme", in June, 2018, the number of monthly page views reached 310 million, the number of members reached 5 million, and the total number of reviews was 14 million[8]. Numerous women exchange information about beauty and cosmetics through the service of @cosme. Hence, users can compare cosmetic items of various cosmetics brands. Reviews consist of a review text, scores and tags regarding effects, etc.

Furthermore, the system stores profile data that include information about age and skin type. Therefore, users who wish to browse the reviews can search the reviews according to their own desired characteristics, for example, reviews can be sorted by scores or filtered for one effect.

Several studies have been conducted on review analysis owing to the spread of web sites that share review information. For example, Hiroshi et al. extracted a remark about product reputation from enormous texts with descriptive sentences such as the questionnaire, and checked the writer's intention[9].

In our previous study, we analyzed reviews of cosmetic items and constructed an evaluation expression dictionary by analyzing reviews. In addition, we developed a method that can automatically calculate the score of each aspect of a "Face lotion" based on the dictionary[2].

O'Donovan et al. evaluated their AuctionRules algorithm, which is a dictionary-based scoring mechanism for eBay reviews of Egyptian antiques[10]. They showed that the approach was scalable and more specifically, that a small amount of domain knowledge could greatly improve the prediction accuracy compared against traditional instance-based learning approaches.

Titov et al. proposed a statistical model for sentiment summarization[11], which was a joint model of text and aspect ratings. To discover the corresponding topics, this model used aspect ratings; thus, it was able to extract textual evidence from reviews without the need of annotated data.

Pham et al. proposed a new method based on the least-squares model using both known aspect ratings and the overall rating of reviews to identify the overall aspect weights directly from numerous consumer reviews[12]. This method estimated the score of all aspects but there is not content about the evaluation viewpoint within a review.

As previously stated, there are several studies in which reviews have been analyzed. However, no study exists in which efforts have been made to develop a method for the automatic scoring of review texts and for proposing tags for cosmetic items.

3. Evaluation Expression Dictionary for Cosmetic Review Recommendation

In this section, we provide an overview of our cosmetic review recommender system and the method used to construct a evaluation expression dictionary.

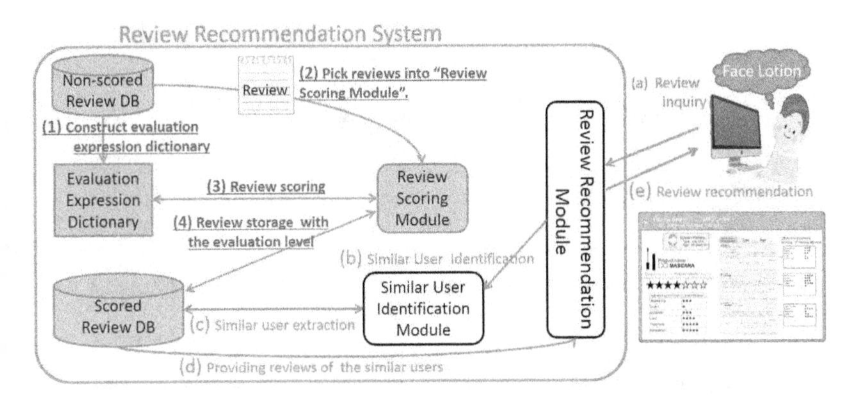

Fig. 1. Conceptual diagram of the cosmetic item review recommender system

Dictionary

Score	Degree	Feature	Keyword
5	considerably		Moisten
7	Very much		Moisten
	:		
2		issues	skin irritation
	:		

A review of Face Lotion A

Review by user X
Skin is considerably moistened when I use lotion A and it is moistened very much! But it may cause skin irritation issues···

Estimated Score for Face Lotion A

Aspect	Estimated Score
Moisturizing	★★★★★★
Hypoallergenic	★★
:	:

←(5+7)/2=6points
←2points

Fig. 2. Review scoring method based on the evaluation expression dictionary

3.1. *Overview of the cosmetic review recommender system*

The aim of the cosmetic review recommender system is to provide recommendations of truly useful reviews for each target user. Fig. 1 shows a conceptual diagram of the cosmetic item and review recommender system, which is the final goal of our research. In Fig. 1, the blue numbers from (1) through (4) correspond to the automatic scoring process of the cosmetics review. The red Roman numerals from (i) through (v) correspond to the review recommendation process. More detailed procedure of each process

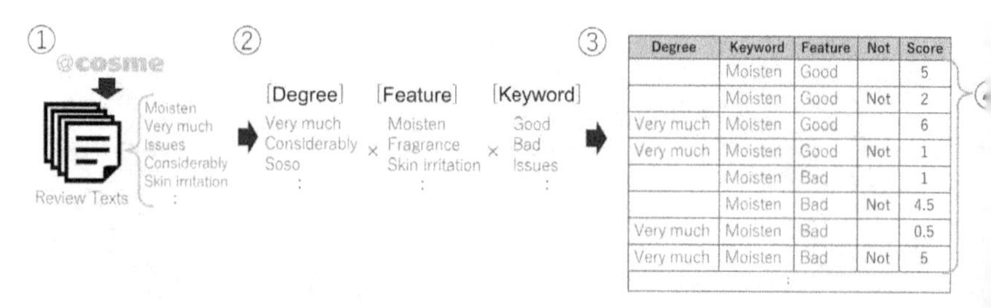

Fig. 3. Constructing the Co-occurrence Keyword-based Evaluation Expression Dictionary

are shown below:

Cosmetic Review Automatic Scoring Process

(1) Construct an evaluation expression dictionary which includes pairs of evaluation expression and its score by analyzing reviews sampled from non-scored DB.
(2) Pick up reviews from non-scored DB to score them.
(3) Automatically score reviews picked up in step (2) based on the evaluation expression dictionary constructed in step (1).
(4) Put reviews scored in step (3) into a scored review DB.

Review Recommendation Process

(a) User gives the name of a cosmetic item that he or she is interested in.
(b) System Refers to the "similar user identification module" in order to extract similar users to the target user of step (a).
(c) "Similar user identification module" obtains the information of reviews and reviewers and identify similar users to the target user.
(d) Provide reviews of similar users identified in step (c) to the "Review recommendation module".
(e) System recommends suitable reviews to the target user.

Our system calculates the scores for various aspects to judge the similarity of users and to recommend cosmetic reviews. Fig. 2 shows the review scoring method for non-scored reviews based on the evaluation expression dictionary. The system reads a non-scored review text and identifies evaluation expressions existing in the reviews. Afterwards, the system gives a score to each evaluation expression according to the dictionary. In this step, if there are several expressions for one aspect, the system calculates the average score for the aspect(Fig. 2). For example, in Fig. 2, two expressions

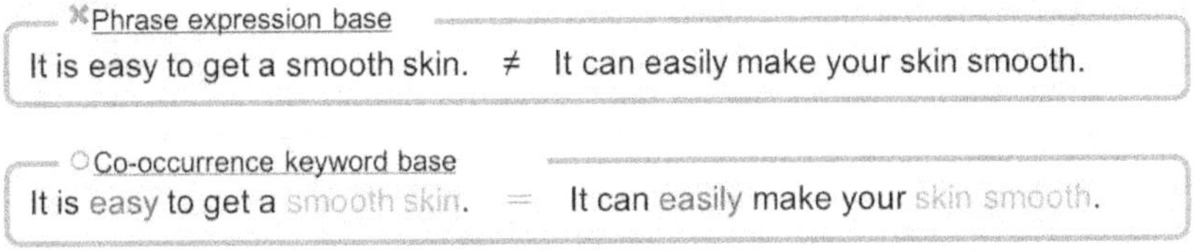

※Phrase expression base

It is easy to get a smooth skin. ≠ It can easily make your skin smooth.

○Co-occurrence keyword base

It is easy to get a smooth skin. = It can easily make your skin smooth.

Fig. 4. Differences in detecting the evaluation expression between the Phrase expression base and the Co-occurrence keyword base

for "moisturizing aspect" exist in the review and our system calculates the average score as the score of "moisturizing aspect".

3.2. *Constructing the evaluation expression dictionary*

Fig. 3 shows the method used to construct the evaluation expression dictionary. To construct the evaluation expression dictionary, we utilized review data provided by @cosme. First, we analyzed phrasal evaluation expressions extracted from reviews. Second, we divided the phrasal expressions into aspect keywords, feature words and degree words. Finally, we constructed the dictionary by assembling their co-occurrence relations and the evaluation scores. Overall, there are several types of cosmetic items. As a first step, we sought to construct an evaluation expression dictionary of the "Face lotion" aspect because "Face lotion" is used by numerous people. In addition, various evaluations may exist for only one "Face lotion" product owing to differences in the skin types of the users. As shown in Fig. 4, the phrases "It is easy to get a smooth skin" and "It can easily make your skin smooth" are semantically nearly identical; however, as a phrase, each of them is different from one another. Hence, it may be possible to detect more evaluation expressions based on the co-occurrence keyword-base dictionary than based on the phrase-based dictionary. Therefore, we use the co-occurrence keyword-base dictionary.

By using this procedure, we developed evaluation expression dictionary for "face lotion" manually[2]. This dictionary stores 1,332 evaluation expressions for "face lotion". Then, in order to verify the effectiveness of our proposed method, we evaluated the results of the automatic scoring based on the dictionary. As the results, our system scores with around 81% accuracy. However, as mentioned in Section 1, it is not a good idea to develop the dictionaries manually because of it takes lot of man-hours.

Table 2. The number of reviews and frequent appearing words for each cosmetic category

Category	Cleansing	Face wash	Face lotion	Emulsion	Lips	Blush
number of reviews	42	40	41	40	38	40
number of frequent appearing words	45	51	43	47	43	42

4. Feature Analysis of Evaluation Expression for Each Category of Cosmetic Items

In this section, we describe the features analysis of evaluation expressions for each cosmetic category. In order to construct an evaluation expression dictionary efficiently, we analyze the relationship between the purpose of use and the evaluation expressions.

4.1. *Target datasets for the analysis*

We already constructed an evaluation expression dictionary for "face lotion". Aiming to construct new dictionaries efficiently by using already constructed dictionary, we analyzed the evaluation expressions of "emulsion" which is the same intended use of "face lotion" like "moisturizing". Furthermore, we analyze the "cleansing and face wash". Intended use of "cleansing and face wash" is different from "face lotion", but the large cosmetic category is the same (skin care). Thus, we consider the possibility of expansion of the dictionary for "face lotion" to the "skin care" category. In order to expand the dictionaries to all the cosmetic categories, we analyze the "Lips" and "Blush" which are coloring purpose items. According to these analysis, we will discuss an automatic construction policy of evaluation expression dictionary for all cosmetic categories.

For this analysis, we use review texts of over 50 characters for each cosmetic categories. We gather these 241 review texts, then extract characteristic expressions for each category by focusing on the frequent appearance of words of verb, adjective, noun, and adverb. Detailed information of target review data are shown in TABLE 2.

4.2. *Results of the analysis*

In order to analyze the features of each cosmetic category, we use two cases (CASE 1, CASE 2) to calculate the similarity of frequently appearing words. For calculating the similarity, we adopt the cosine similarity metric.

<u>CASE 1</u>

First, we create 129 dimensional feature vectors based on the frequency

Table 3. Part of appearance count for frequently occurrence words for each cosmetic category

	Cleansing	Face wash	Face lotion	Emulsion	Lips	Blush
Freshen	3	7	4	4	0	0
Pearl	0	0	0	0	1	5
Dry	6	7	4	9	3	5
Mat	0	0	0	0	1	2
:	:	:	:	:	:	:

Table 4. CASE 1 : The results of cosine similarity between different cosmetic categories

	Cleansing	Face wash	Face lotion	Emulsion	Lips	Blush
Cleansing	-	0.51	0.18	0.23	0.25	0.19
Face wash	0.51	-	0.28	0.31	0.08	0.08
Face lotion	0.18	0.28	-	0.84	0.09	0.12
Emulsion	0.23	0.31	0.84	-	0.11	0.14
Lips	0.25	0.08	0.09	0.11	-	0.86
Blush	0.19	0.08	0.12	0.14	0.86	-

Table 5. Part of the presence or absence(0 or 1) of frequently occurring words for each cosmetic category

	Cleansing	Face wash	Face lotion	Emulsion	Lips	Blush
Freshen	1	1	1	1	0	0
Pearl	0	0	0	0	1	1
Dry	1	1	1	1	1	1
Mat	0	0	0	0	1	1
:	:	:	:	:	:	:

of appearing words extracted from all review texts. Then, we calculate the similarities between each cosmetic category. TABLE 3 shows a part of the appearance count for frequently occurrence words of each cosmetic category. And TABLE 4 shows the results of cosine similarity between different cosmetic categories.

According to the results of CASE 1, we can find the cosmetic categories that have the most similar feature vectors of frequently appearing words. Thus, we can estimate that categories, such as "Cleansing and Face wash", "Face lotion and Emulsion", and "Lips and Blush", are similar cosmetic categories that contain similar evaluation expressions. By using the analysis method in CASE 1, we can construct the dictionaries for similar purpose categories.

Table 6. CASE 2 : The results of cosine similarity between different cosmetic categories

	Cleansing	Face wash	Face lotion	Emulsion	Lips	Blush
Cleansing	-	0.58	0.41	0.48	0.34	0.32
Face wash	0.58	-	0.51	0.47	0.28	0.28
Face lotion	0.41	0.51	-	0.65	0.35	0.28
Emulsion	0.48	0.47	0.65	-	0.38	0.34
Lips	0.34	0.28	0.35	0.38	-	0.49
Blush	0.32	0.28	0.28	0.34	0.49	-

Table 7. Concordance rates of frequent words between different cosmetic categories

	Cleansing	Face wash	Face lotion	Emulsion	Lips	Blush
Cleansing	-	55%	42%	47%	35%	33%
Face wash	62%	-	56%	49%	30%	31%
Face lotion	40%	47%	-	62%	35%	29%
Emulsion	49%	45%	67%	-	40%	36%
Lips	33%	25%	35%	36%	-	50%
Blush	31%	25%	28%	32%	49%	-

CASE 2

In CASE 2, we create dimensional feature vectors based on the presence or absence (0 or 1) of appearing words extracted from all review texts. Then, we calculate the similarities between each cosmetic category, in order to find similar relationships other than "Cleansing and Face wash", "Face lotion and Emulsion", and "Lips and Blush". TABLE 5 shows a part of the presence or absence (0 or 1) of frequently occurring words for each cosmetic category. TABLE 6 shows the results of cosine similarity between different cosmetic categories in CASE 2. In addition, TABLE 7 shows Concordance rates of frequent words between different cosmetic categories.

The results of CASE 2 indicated that there are many common words in reviews for "Cleansing", "Face wash", "Face lotion" and "Emulsion". Thus, we may say that our analysis method in CASE 2 can estimate similarities between different cosmetic categories.

4.3. *Construction Policy of Evaluation Expression Dictionaries*

TABLE 8 shows the frequent occurrence words in the high similarity categories of the four skin care categories, such as "Cleansing and Face wash", and "Face lotion and Emulsion" category. We discuss the construction pol-

Table 8. Frequent words of "skincare"

Category	Cleansing, Face wash, Face lotion, Emulsion	Cleansing, Face wash	Face lotion, Emulsion
Frequent words	Moisturizing, Refresh, Allergies, Aging, Cheap, Value, Smooth, Dry, Expensive, Stimulation, Moisture, Price, Rough skin	Clear, Smooth, Dirty, Wash, Moderate, Washing, Tautness, Transparency, Buying, Foam, Removing	Refreshing, Thickening, Plump, Rough, High price, Penetration, Sensitive, Wrinkle

Table 9. Frequent words of "lips", "blish" and "skincare"

Category	cleansing, face wash, face lotion, lips, blush	lips, blush
Frequent words	Moisturizing, Dry, Cheap, Value	Lustrous, Mat, Pearl, Lame, Blood, Color, Coloring, Coloring, Shades, Petit Price

icy of evaluation expression dictionaries based on this result. We aim to share the dictionary between categories as much expressions as possible .

The expressions, "moisture", "freshen", "hypoallergenic", "rough skin", "aging care", and "cost-performance" occurred commonly with four "Skin care" categories. The expressions that related to "Foam" and "Tautness" occurred commonly with "Cleansing and Face wash" categories. The expressions that related to "Refreshing or Thickening" occurred commonly to "Face lotion and Emulsion" categories. According to above results, we can find the evaluation expressions that can be shared among the categories (Fig.5). We can also find this fact from the concordance rate of frequency words between categories shown in TABLE 7.

TABLE 9 shows the frequently occurring words in the high similarity categories between "Lips", "Blush", and "Skincare" categories. The expressions that related to "Moisturizing", "Cost-performance" occurred commonly in all six categories (Lips, Blush, and Skin care). The expressions related to "coloring", "Color holding", "lame/pearl", and "shiny or matte" occurred commonly to the "Lips and Blush" categories. According to the common words described above, we can find the evaluation expressions that can be shared among the categories as shown in (Fig. 6).

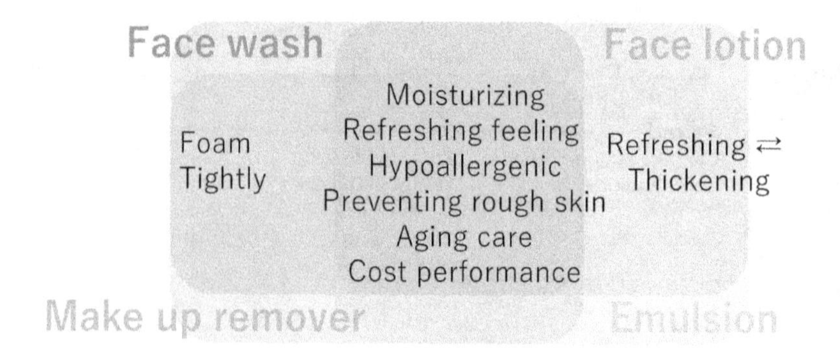

Fig. 5. Common words in the "Skin care" categories

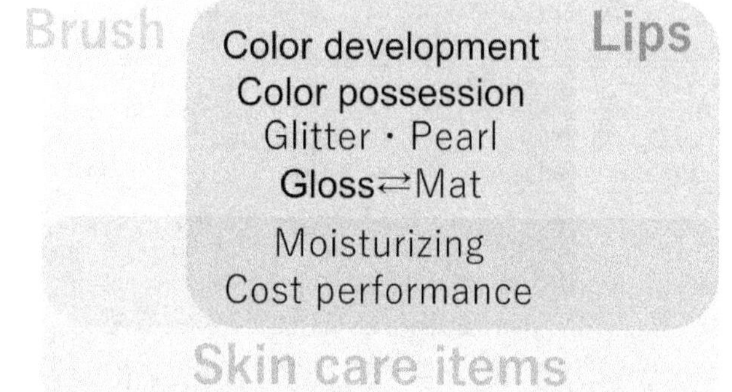

Fig. 6. Common words in the "Lips", "Blush", and "Skin care" categories

Consequently, the results and the considerations given above indicate that there are both similar evaluation expressions and not similar evaluation expressions between reviews for skincare items and reviews for makeup items. This indicates that we can utilize common evaluation expressions and aspects among different cosmetic categories in order to construct each evaluation expression dictionary.

4.4. *Applying evaluation expression dictionary from existing items to similar items*

After analyzing the characteristics of the different cosmetic categories, we found that two items in particular, "Lotions" and "Emulsions" are highly

Table 10. Evaluation rate and MAE (Mean Absolute Error)

	Evaluation rate	MAE (Mean Absolute Error)
Face lotion	0.73	1.42
Emulsion	0.65	1.93

similar items as they contain a large number of evaluation expressions and evaluation items which are similar. Therefore, it would be possible to apply the evaluation expression dictionary that has been developed in this study to these two cosmetic categories to evaluate our approach. We extracted 30 random reviews and performed automatic scoring for each item category using our approach and the percentage of the categories that were scored correctly was used to calculate the evaluation rate. In addition, the average absolute error between the scores given manually by the users and the scores obtained using the automatic scoring approach based on our evaluation expression dictionary was calculated for each evaluation item.

TABLE 10 shows the evaluate rate and the average absolute error for the cosmetic items in the skin lotion and emulsion categories. The evaluation rate was 0.73 for the skin lotion items and 0.65 for the emulsion items. The average absolute error was also quite similar for the skin lotion items (1.42) and the emulsion items (1.93). The fact that the evaluation expression dictionary was constructed from user reviews of products in the skin lotion category could explain the higher evaluation rate and lower average absolute error score for skin lotion items. This result shows that the existing evaluation expression dictionary could be used to evaluate other item categories with high similarity and highlights the effectiveness of our dictionary.

5. Conclusion

The aim of our research is to develop a recommender system for cosmetic items and reviews to help consumers. In order to realize such a recommender system, we proposed and discussed how to develop a method for automatic construction of evaluation expression dictionary for all categories by analyzing the tendency of both similar evaluation expressions and dissimilar evaluation expressions between reviews for skincare items and reviews for makeup items.

As the result, we recognized that evaluation expressions differ depending on their cosmetic item categories. Moreover, it was found that we can utilize common evaluation expressions and aspects among different cosmetic categories in order to construct each evaluation expression dictionary.

A further direction of this study will be to develop evaluation expression dictionaries for all cosmetic item categories and to work through developing a recommender system for cosmetic items and reviews.

Acknowledgment

We would like to express our gratitude to istyle Inc. who provided review data for cosmetic items, and Prof. Panote Siriaraya of Kyoto Institute of Technology who helped us to implement our system.

References

1. @cosme, `http://www.cosme.net/`, (Accessed on 8th January 2019)
2. Yuki Matsunami, Mayumi Ueda, Shinsuke Nakajima, Takeru Hashikami, John O'Donovan, ByungKyu Kang, "Mining attribute-specific ratings from reviews of cosmetic products", in *Transactions on Engineering Techonologies(International MultiConference of Engineers and Computer Scientists 2016)*, pp.101-114, Springer, 2017.
3. Asami Okuda, Yuki Matsunami, Mayumi Ueda, and Shinsuke Nakajima, "Finding similar users based on their preferences against cosmetic item clusters",in *Proceedings of the 19th International Conference on Information Integration and Web-based Applications & Services*, pp.156-160, 2017.
4. Mayumi Ueda, Yuki Matsunami, Panote Siriaraya, and Shinsuke Nakajima, "Developing Evaluation Expression Dictionaries for the Cosmetic Review Recommendation," *Lecture Notes in Engineering and Computer Science: Proceedings of The International MultiConference of Engineers and Computer Scientists 2019*, 13-15 March, 2019, Hong Kong, pp236-241.
5. Amazon.com, `http://www.amazon.com/`, (Accessed on 8th January 2019)
6. Priceprice.com, `http://ph.priceprice.com/`, (Accessed on 8th January 2019)
7. Byunkyu Kang, et al., "Inspection Mechanisms for Community-based Content Discovery in Microblogs" *IntRS'15 Joint Workshop on Interfaces and Human Decision Making for Recommender Systems at ACM Rccommender Systems.*
8. The site data of @cosme (May.2017), istyle Inc., `http://www.istyle.`

`co.jp/business/uploads/sitedata.pdf` (in Japanese), (Accessed on 8th January 2019)

9. Hiroshi Kanayama, et al., "Textual demand analysis: detection of users' wants and needs from opinions." Proceedings of the 22nd International Conference on Computational Linguistics-Volume 1. Association for Computational Linguistics, 2008.
10. John O'Donovan, et al., "Extracting and Visualizing Trust Relationships from Online Auction Feedback Comments.," *International Joint Conference on Artificial Intelligence (IJCAI'07), 2007.*
11. Ivan Titov et al., "A Joint Model of Text and Aspect Ratings for Sentiment Summarization," *46th Meeting of Association for Computational Linguistics (ACL-08)*, Columbus, USA, pp.308-316, 2008.
12. Duc-Hong Pham, et al., "A least square based model for rating aspects and dentifying important aspects on review text data," *the 2nd National Foundation for Science and Technology Development Conference on Information and Computer Science*, pp.265-270, 2015.

A Web Search Method Based on Complaint Data Analysis for Collecting Advice

Liang Yang
Kwansei Gakuin University 2-1 Gakuen,
Sanda-shi, Hyogo, 669-1337, Japan
E-mail: dui93794@kwansei.ac.jp

Daisuke Kitayama
Kogakuin University 1-24-2 Nishi-Shinjuku,
Shinjuku-ku, Tokyo, 163-8677, Japan
E-mail: kitayama@cc.kogakuin.ac.jp

Kazutoshi Sumiya
Kwansei Gakuin University 2-1 Gakuen,
Sanda-shi, Hyogo, 669-1337, Japan
E-mail: sumiya@kwansei.ac.jp

Nowadays, there are a large number of users who post complaints about a certain service on the Internet. Because users have various values and views, even if they receive the same service, they may complain in different ways. However, it is quite difficult to respond to various user demands for service in real time and there are almost no direct solutions when users feel dissatisfied with a certain service. Therefore, in this paper, we propose a web search method by analyzing complaint data from Fuman Kaitori Center. First, the system generates query keywords according to various user complaints about a certain service by calculating the score of each query. Then suitable web pages containing advice are recommended from the results of the query. This advice could address users' dissatisfaction and respond to their various demands in a comprehensive way. Also, we verify the usability of proposed system by using a questionnaire survey evaluation.

Keywords: Web Search, Query Extraction, Advice, Complaint Data.

1. Introduction

In recent years, many users post negative reviews about a certain service online. However, it is quite difficult to respond to various user demands for service in real time as the service is provided by the company. In addition, there are almost no direct solutions when users feel dissatisfied with a certain service. Therefore, this paper is focused on user complaints related to services and proposed a system to search for advice that could address users' dissatisfaction by generating query keywords from complaint reviews [11]. This advice contains merits of the service users may not be aware of and could respond to their different demands in a comprehensive way. An example of advice recommendation is described in Figure 1.

The remainder of this paper is structured as follows. Section 2 presents a brief summary of related work. Section 3 introduces the dataset we use for research and explains the proposed system. Section 4 discusses the experimental results and the evaluation of the proposed system. Finally, Section 6 concludes this paper and discusses future work. LATEX commands as much as possible.

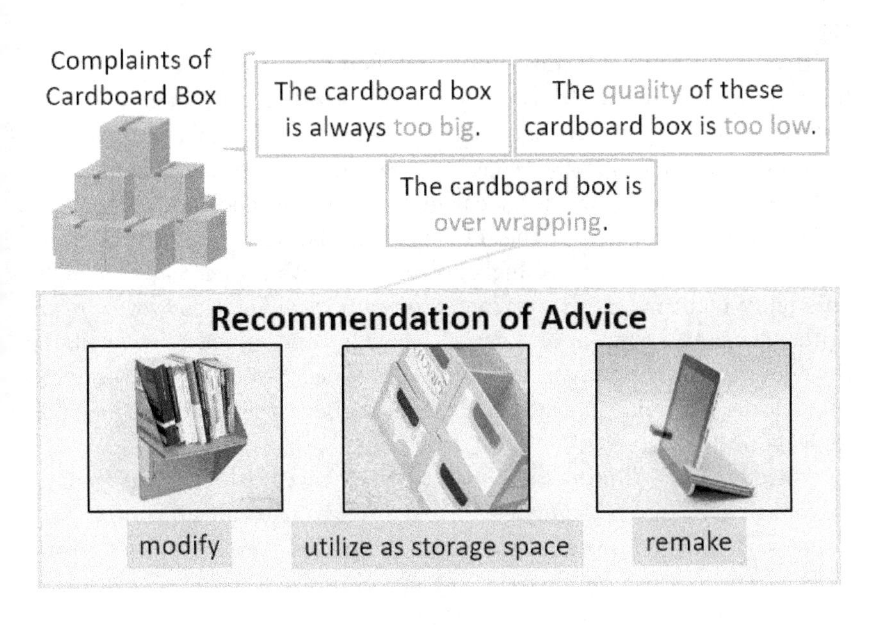

Fig. 1. Example of Advice Recommendation

2. Related Work

2.1. *FKC Dataset*

The FKC dataset has been used for several studies in recent years. Mitsuzawa et al. [1] presented the FKC dataset which is from Fuman Kaitori

Center (FKC). "Fuman" means dissatisfaction in Japanese. The FKC is a Japanese consumers' negative opinion data collection and analysis service. In our work, we used and analyzed the FKC dataset.

Hasegawa et al. [2] analyzed and visualized the contents of the FKC dataset such as the distribution of users' ages, jobs, and gender. In our work, we determined the target of the experiment based on their results.

2.2. *Topic Word Extraction*

Sakai et al. [5] proposed a method to extract negative words as the expressions of dissatisfaction from blogs. They extracted nouns, adjectives to make a dissatisfaction expression dictionary. In our work, we only extract nouns because nouns can explain and represent the content of users' complaints.

Hashimoto et al. [6] proposed a method to extract important topics from newspaper and detect social problems based on document clustering. Ustumi et al. [4] proposed a method to extract technological solutions to social problems such as medical issues from the news. They extracted technological solution words by calculating the relevance of problems and technologies. They defined the relevance calculation as problem relevancy and technical relevancy. A higher value of relevancy indicated a higher possibility of being able to extract a technological solution word. In our work, we use this concept and extract the advice topic word by calculating the relevance of the company and complaint topic. However, we hypothesize that a lower relevancy indicates a higher probability that a word is an advice topic word.

Yoshida et al. [7] proposed a method to extract features terms from the customer reviews of e-commerce sites in order to recommend similar items to users. They used polarity analysis to calculate the degree of importance of feature words by counting the number of positive reviews, negative reviews, and positive ratings. In our work, we also use polarity analysis to evaluate advice topic words. In addition, we weight words according to the result of polarity analysis.

2.3. *Query Generation*

Song et al. [9] and Kajinami et al. [10] proposed a system to generate query keywords that can support a user's search intention. Kakimoto et al. [8] proposed a system to extract query keywords from the closed caption data of TV programs to recommend web pages related to tourism and events based on users' preferences. In our work, we extract query keywords from negative reviews to recommend web pages of advice with the aim of addressing a user's dissatisfaction with a certain service.

2.4. *Recommender System using Complaint Data*

Hayashi et al. [3] proposed a system to recommend appropriate products for users according to their complaints. This system could directly resolve users' dissatisfaction by recommending certain substitute product. In our work, we proposed a system to resolve user complaints about services instead of products in an indirect way.

3. Proposed Method

3.1. *Overview*

In this paper, we propose a web search method by analyzing complaint data from Fuman Kaitori Center for recommendation of advice in order to address users' dissatisfaction about a certain service. Figure 2 shows the system flow of our proposed method. First, we extract the company name and complaint topic words by calculating the importance of the nouns in the negative reviews. Second, we obtain candidate search keywords of these extracted words. Then, the system extracts the advice topic word by calculating the relevancy of the candidate keywords to the FKC dataset and score them using morphological and polarity analyses. Third, we create the query by combining the company name, complaint topic word, and advice topic word according to their various complaints about a certain service. Finally, suitable web pages containing advice that could address a user's complaint are recommended from the results of the query.

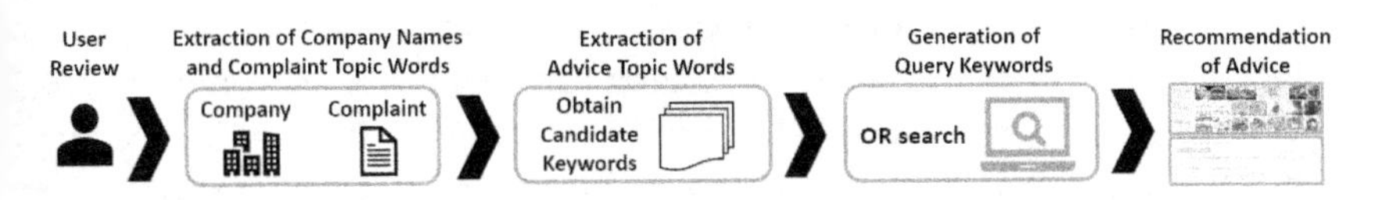

Fig. 2. System Overview

3.2. *Dataset*

In this study, we analyze a dataset of complaints from the Fuman Kaitori Center, which is provided by Insight Tech Inc. from the National Institute of Informatics. In this paper, we refer to the Fuman Kaitori Center's dataset as the FKC dataset. The Fuman Kaitori Center is a website on which users can post their complaints about topics such as products, services, education, work, and relationships. Moreover, users get points when they post complaints that they can exchange for coupons for online shopping websites. This dataset contains about 5 million negative reviews that were posted from 18 March 2015 to 12 March 2017 by around 100,000 users. Each negative review contains the information shown in Table 1. In FKC dataset, each category contains several subcategories, and each subcategory contains several companies.In this paper, because we focus on user service complaints, our proposed system uses the data fields for "company" and "text" .

Table 1. Data Structure of the FKC Dataset.

Data Item	Content
post_id	complaint ID
user_id	Fumankaitori Center ID
category	complaint category
sub_category	detailed complaint category
company	company name
product	product name
text	negative review

3.3. *Extraction of Company Names and Complaint Topic Words*

Our proposed system extracts company names from FKC dataset directly from the company field of each record. Next, we extract the complaint topic word by analyzing negative reviews. In this paper, we only use the negative reviews that are labeled with the company name. To extract complaint topic words, we first extract all companies' negative reviews for one subcategory and extract all nouns from the negative reviews. Next, we calculate the importance of each noun using the following equation.

$$\frac{tf}{|A|} \times \frac{tf}{\sum_{d \in D} tf_d} \tag{1}$$

Here, tf is defined as the number of occurrences of a particular noun in the complaints for a certain company, |A|is defined as the number of all nouns in the complaints for a certain company, and $\sum_{d \in D} tf_d$ is defined as the number of occurrences of certain noun for the complaints for all companies. Finally, we extract all nouns whose importance values are above the determined threshold value and define them as that company's complaint topic words.

3.4. *Extraction of Advice Topic Words*

To extract the advice topic word, we first obtain candidate search keywords of the company name and complaint topic word. Because the FKC dataset is full of negative reviews, we hypothesize that candidate keyword that are less relevant to the FKC dataset will make better advice topic words. To verify this hypothesis, we calculated the relevance of these candidate keywords for each company and each complaint topic word and define as "company relevancy" and "complaint topic relevancy". It is calculated using the following equation.

$$company\, relevancy = \frac{R_{cd}}{R_c} \tag{2}$$

$$complaint\, topic\, relevancy = \frac{R_{td}}{R_t} \tag{3}$$

Here, R_{cd} is defined as the number of occurrences of certain candidate keyword in complaints for the company in the FKC dataset and R_c is defined as the number of negative reviews of that company. R_{td} is defined as the number of occurrences of the candidate keyword with the complaint topic word in the negative reviews of the FKC dataset and R_t is defined as the number of negative reviews with that complaint topic word.

After that, to exclude some negative words as well as verbs and adjectives which do not help users acquire advice, we weight candidate keywords using morphological and polarity analyses, as shown in Table 2.

Finally, we calculate the final score of the candidate keywords by combining the arithmetic mean of the company and complaint topic relevancies with the weight as the following equation.

$$Score = \frac{relevancies}{2} \times Weight \tag{4}$$

After calculating the final score of each candidate keyword, we determine the threshold value for each company. Candidate keywords those scores are under the threshold value become the advice topic words.

Table 2. Weight for Candidate Keywords.

Result of Analysis	Weight
negative	0.8
verb	0.7
adjective	0.7
proper noun(place name)	0.7
proper noun(organization name)	0.3
common noun	0.3
verbal noun	0.1

3.5. *Generation of Web Search Queries*

In this study, each company name and complaint topic word are matched with several advice topic words. To search for suitable websites, We use an OR-based search method to acquire advice websites. Our proposed system generates the query based on one company name, one complaint topic word, and one advice topic word.

3.6. *Recommendation of Advice*

Our proposed system recommends suitable web pages containing advice from the results of the query which is based on users' complaints. Figure 3 shows the user interface of our proposed system. First, the system generates several queries by analyzing user's negative review. Next, user can choose and browse the web page based on their needs by a web search using the offered queries. The system recommends the advice information that could address user's dissatisfaction expressed in the negative review.

Fig. 3. User Interface

4. Experiment and Evaluation

4.1. *Experiment*

In this study, we conducted an experiment to extract the complaint and advice topic words in order to verify the feasibility of proposed system. For this experiment, we analyzed the subcategory of "IT web services" of the FKC dataset, which is under the category "industry." We analyzed 1,000 negative reviews for each of three companies.

First, we extracted the complaint topic words and determined different threshold values for each of the three companies. For company A, we extracted 186 complaint topic words above the threshold value of 0.00080. For company B, we extracted 144 complaint topic words above the threshold value of 0.00076. For company C, we extracted 86 complaint topic words above the threshold value 0.00080. Table 3 shows examples of the complaint topic words for each company. These examples show that each complaint topic word implies the object of different users' dissatisfaction.

Table 3. Example Complaint Topic Words.

Company	Complaint Topic Words
A	purchase, prime, delivery, review,delivery fee, order, membership, return, post, gift, cardboard box, price, sign, yamato, book
B	news, question, premium,answer, auction, article, title, navigation, mail, shopping, ID, weather forecast, transaction, search, comment
C	purchase, prime, delivery, review,delivery fee, order, membership, return, post, gift, cardboard box, price, sign, yamato, book

Next,we extracted advice topic words from the candidate keywords that had a score less than 0.0043, 0.0020, and 0.0033 for companies A, B, and C, respectively. Some examples of these words are shown in Table 4. As TABLE 4 shows, the proposed method is sufficient for ranking candidate keywords.

4.2. *Evaluation*

In this paper, we conducted a questionnaire-based survey to evaluate the usability and effectiveness of the proposed method.The questionnaire-based

Table 4. Example Advice Topic Words.

Company	Complaint Topic Words	Advice Topic Words
A	point	charge, present, how to save up, how to use, credit card
B	setting	security, group friend, privacy, initialization, recommendation
C	premium	privilege, merit cancellation of agreement, magazine

survey contained following 3 questions. For Q1, we extracted all nouns from the negative review for respondents to choose from. For Q2, we provided 15 queries for each negative review to choose from. 5 of the queries' candidate keywords were made by ourselves. For Q3, we evaluated the satisfaction of the result of advice recommendation.
Q1:Please choose one word which you think could represent the dissatisfaction of the following reviews.
Q2:please choose the query that you think the contents returned by a search using this query keyword could address the complaints found in the negative review.(multiple choices are allowed)
Q3:Please make a web search with the queries you have chosen. Do you feel satisfied with the contents of the advice?

4.3. *Result and Discussion*

We collected the answers of 10 respondents, and the results are shown in Table 5 and Table 6. We defined those nouns and queries were chosen by over 5 answers as true positive.

Table 5. Result of Q1.

	p@1	p@3	p@5
Average of p@k	0.60	0.38	0.33
Average of r@k	0.50	0.63	0.83
F-measure	0.55	0.47	0.48

The result of Q1 showed that if we search by using those nouns are with the highest value of importance for one time only, 60% of appropriate complaint topic word can be extracted from the negative review. It not only shows that the proposed method is effective, but also explains the method to rank nouns by calculating the importance performed well.

Table 6. Result of Q2.

	p@1	p@3	p@5
Average of p@k	0.70	0.67	0.60
Average of r@k	0.18	0.50	0.75
F-measure	0.28	0.57	0.67

The result of Q2 showed that if we search advice by using the query which is with the lowest score for one time only, 70% of appropriate queries can be offered to make web search in order to address the complaints expresses in the negative reviews. This result demonstrated that scoring candidate keyword is effective. However, we found out that the longer the candidate keyword was, the lower the score will be when making the candidate keywords . In the future, we plan to develop a method to ensure if the candidate keyword is related to the complaint topic word to better exclude noise in the results.

For Q3, the result showed that 75% of the answer felt satisfied with the contents of advice they searched with the queries they've chosen. From this result, it is observed that by using the proposed system could address users' dissatisfaction and the recommendation of advice respond to the demands of different users. Moreover, it implied by using proposed system can help users to release their burden when searching for advice comparing to traditional search engine.

5. Conclusion

In this paper, we proposed a web search method by analyzing complaint data to recommend suitable advice. We extracted query keywords from various user complaints about a certain service by calculating the score of each query. Then suitable web pages containing advice are recommended from the results of the query. In addition, we evaluated the effectiveness and usability of the proposed system through a questionnaire survey, and the results shows that the generated query keywords would be useful for collecting advice. In addition, the recommendation of advice returned by query keywords could address users' dissatisfaction with a service and respond to different user demands in a comprehensive way.

In the future, we plan to evaluate the satisfaction of each query and analyze the result. Furthermore, we will consider new methods to obtain candidate query keywords which users are hard to associate to enhance the usability of the proposed system.

Acknowledgment

In this paper, we used FKC Data Set provided for research purposes by National Institute of Informatics in cooperation with Insight Tech Inc.

References

1. Kensuke Mitsuzawa, Maito Tauchi, Mathieu Domoulin, Masanori Nakashima and Tomoya Mizumoto. " FKC Corpus: a Japanese Corpus from New Opinion Survey Service," *In proceedings of the Novel Incentives for Collecting Data and Annotation from People: types, implementation, tasking requirements, workflow and results, Portoro, Slovenia* pp.11-18, May. 2016
2. Tooru Hasegawa and Daisuke Kitayama. "The Visualization of Dissatisfaction Groups using Dissatisfaction Dataset," *DEIM Forum 2017*, P7-1. (In Japanese)
3. Toshinori Hayashi, Yuanyuan Wang, Yukiko Kawai, and Kazutoshi Sumiya. "An E-Commerce Recommender System using Complaint Data and Review Data," *Proc. of ACM IUI2018 Workshop on Web Intelligence and Interaction (WII 2018).*
4. Kazuo Utsumi, Takashi Inui, Taiichi Hashimoto, Koji Murakami and Masamichi Ishikawa. "Extraction of Critical Knowledge concerning Social Problems and their Technological Solutions," *Socio Technology Research Journal*, vol. 6, pp. 187–198, Mar. 2009.
5. T. Sakai and Ko Fujimura. "Discovering Latent Solutions from Expressions of Dissatisfaction in Blogs," *Information Processing Society of Japan*, vol. 52, no. 12, pp. 3806–3816, Dec. 2011.
6. Taiichi Hashimoto, Koji Murakami, Takashi Inui, Kazuo Utsumi and Masamichi Ishikawa. "Topic Extraction and Social Problem Detection based on Document Clustering," *Socio Technology Research Journal*, vol. 5, pp. 216-226, Mar. 2008.
7. Tomoshi Yoshida and Daisuke Kitayama. "An Evaluation of Feature

Term Extraction Method based on Polarity Analysis from Customer Reviews," *IEICE-DE2016-5* vol. 116, no. 105, pp. 19–24, Jun. 2016.
8. Honoka Kakimoto, Toshinori Hayashi, Yuanyuan Wang, Yukiko Kawai, and Kazutoshi Sumiya. "Query Keyword Extraction from Video Caption Data based on Spatio-Temporal Features ," *Lecture Notes in Engineering and Computer Science: Proceedings of the International MultiConference of Engineers and Computer Scientists 2018*, pp. 405–408, Mar.2018
9. Ximei Song and Masao Takaku. "Study on Navigation support System based on Users Search Intents," *ARG WI2*, no. 9, 2016.
10. Tomoki Kajinami, Toshiyuki Ogasawara, Jhoji Komiya and Yasufumi Takama. "Application of Keyword Map to Decision Support through Exploratory Search, " *2008 IEEE International Conference on Systems, Man and Cybernetics*, 2177-2181, 2008
11. Liang Yang, Daisuke Kitayama, and Kazutoshi Sumiya."Query Keyword Extraction from Complaint Data for Collecting Advice," *Lecture Notes in Engineering and Computer Science: Proceedings of the International MultiConference of Engineers and Computer Scientists 2019*, pp. 347–351, Mar.2019

LOD-Based Semantic Web for Indonesian Cultural Objects

Gloria Virginia and Budi Susanto
Informatics Department, Universitas Kristen Duta Wacana,
Yogyakarta, 55224, Indonesia
E-mail: virginia@staff.ukdw.ac.id and budsus@staff.ukdw.ac.id
www.ukdw.ac.id

Umi Proboyekti
Information System Department, Universitas Kristen Duta Wacana,
Yogyakarta, 55224, Indonesia
E-mail: othie@staff.ukdw.ac.id
www.ukdw.ac.id

Online media might be a way to introduce as well as preserve cultural heritage. Related to Indonesian culture, there are publication services available which focus only on information sources without metadata properties. This article explains our effort to preserve the Indonesian cultural heritage using semantic web and linked open data for a comprehensive presentation. An OWL-based ontology has been trying to be developed for a repository application consists of 13 ontologies of Indonesian cultural objects based on Description Logic of manually extracted literatures. The development process involved important partners in Special Region of Yogyakarta, i.e. Department of Culture, Regional Library and Archives Agency, Indonesian University Library Forum, as well as cultural practitioners and researchers. The applications of cultural objects are introduced at the end of this article. It is a baby step of a giant dream.

Keywords: Semantic Web; Linked Open Data; Indonesian cultural objects; Repository.

1. Introduction

Indonesia has a rich cultural life supported by more than 1,300 ethnic groups[1]. These cultural assets need to be preserved as a cultural heritage. In 2015, there were 6,238 intangible cultural heritages and 979 cultural heritages spread across 34 provinces of Indonesia[2]. To preserve cultural heritage, in 2017, Republic of Indonesia Government through the Ministry of Education and Culture has recognized 150 new intangible cultural heritages[3] based on 416 proposals from various regions in Indonesia[4]. Therefore, until 2017 there have been 594 intangible cultural heritages managed by Directorate General of Culture, the Ministry of Education and Culture[5].

Tanudirjo[6] emphasized that all parties must play an active role in preserving the richness of Indonesian cultural heritage to have benefit for themselves. Tanudirjo called it the "Cultural Heritage for All" or the "All for Cultural Heritage". Online publication using web infrastructure considered as an effort to introduce and disseminate Indonesia's cultural richness. It has been started by *Sejuta Data Budaya* (SDB) movement through *Perpustakaan Digital Budaya Indonesia*[a] website and *National Library* through *Digital Batavia Online*[b] program. The intention should be improved to the relationship between one cultural object to another. This linkage information is significant by virtue of a cultural object is inseparable from its environment and tends to influence other cultures.

The aim of this research is to develop a Semantic Web-based information catalog of Indonesian cultural objects by generating a knowledge representation model of Indonesian cultural objects and their relationships. The Semantic Web framework[7] supports ontology-based modelling as a form of open knowledge representation and linkages between objects in order to present more representative information. Therefore, the intended online catalog services must support the open data, i.e. Linked Open Data (LOD), characteristic.

This research reinforce the technology framework for an ICT[c] Content enrichment stipulated in 2015-2045 National Research Master Plan of the Research, Technology, and Higher Education Ministry of Republic of Indonesia. Our LOD-based Semantic Web cultural object repository service is possible to benefit various parties with various objectives, e.g. learning media, publication media. Further more, it should encourage the emergence of other relevant application-based service works.

This article explains our effort towards an LOD-based Semantic Web specifically for Indonesian cultural objects, i.e. the methodology and the applications. The narrative starts from describing the state-of-the-art as well as literature study of research related to Semantic Web and Cultural Heritage and is ended by description of consecutive and potential studies.

---

[a]Indonesia Cultural Digital Library. URL: `budaya-indonesia.org`.
[b]`http://bataviadigital.perpusnas.go.id/home/`
[c]Information and Communication Technology

2. Literature Study

Awareness for raising and disseminating information related to the work and objects of Indonesian culture has been conducted by the National Library of Indonesia with the Digital Batavia Online program. Other efforts to provide information services on Indonesian cultural objects are also carried out by the community, e.g. the Indonesian Cultural Digital Library[d].

The Indonesian cultural information service representation framework does not yet follow the open data framework. Based on the Open Knowledge Foundation[8] and W3C[9], every community and government can build a system to collect data and content from anyone. Thus, the application of the open data framework may play a role in increasing the Indonesian Human Development Index (HDI).

Implementation of Semantic Web framework encourages the dissemination of information related to Indonesian cultural objects. The Semantic Web framework provides a repository system that functions as open data and open knowledge for cultural objects. Westrum in Ref. 10 emphasises that the application of open data by including semantic content in the data makes information repository services easy to use and develop. Libraries, museums, and archival institutions function as main information suppliers for the community[11]. The addition of information to the web document will provide a clearer meaning description of the document contents. Providing a description to the meaning of documents can be realized by a machine, through a developed application. This machine understanding feature is based on the meaning representation conveyed by a web document[12].

On the other hand, the application of the Semantic Web to a domain requires the development of knowledge representation models. This knowledge representation model requires an ontology definition. Biffl & Sabou in Ref. 13 stresses that the use of metadata has to be precisely defined in order to reduce the confusion of computer applications that will translate it. There are 2 principles should be followed. First, metadata must describe information in terms that have a clear meaning to the machine and reflect the agreement of the wider community. Second, metadata must be expressed in a representation language that can be parsed and translated by computer applications. Sugumaran & Gulla[14] described the layers in the Semantic Web, i.e. 1) data and metadata; 2) semantics; 3) enabling technology; and 4) environment. Each layer provides a set of functions to

---

[d] `http://budaya-indonesia.org`

perform its layer tasks[15].

The semantics layer provides several functions to fulfil the conditions of well-defined meaning, i.e. ontology language (OWL), rule languages (SWRL/RIF/RuleML), query languages (SPARQL), logic (Description Logic), reasoning mechanisms, and trust. Each data should be represented by standard rules to support data exchange between various applications and systems. At the data layer and metadata, there are Unicode, URI, XML, RDF, and RDF Schema standards.

Ontology is significant for a Semantic Web application to facilitate knowledge sharing and reuse in a distributed system[16]. An ontology that represents static domain knowledge and problem-solving methods, will be used in the Semantic Web services that modeling the reasoning process and deal with the domain knowledge[17]. Grimm, Abecker, and Studer[18] describe the main characteristics associated with the development of ontology in information systems, i.e. 1) interrelation; 2) instantiation; 3) Subsumption; 4) Exclusion; 5) Axiomatization; and 6) Attribution.

Different approaches are proposed for ontology development methodology, i.e. employing frames and first-order logic[19], description logic[20], Unified Modeling Language (UML)[21], and E-R Diagrams to represent an ontology[17]. However, most of the methodologies are based on people's experiences when got involved in ontology development. There are methodologies derived from the practice of ontology development itself, while the other focused on application development in which ontology is created. The example of the first approach are *CYC*[22] (which was based on experience during the development of CYC knowledge base), *Enterprise Ontology*, *Toronto Virtual Enterprise* (TOVE)[23], and *Amaya*[24] (which is produced from ESPRIT KACTUS project). Several methodologies included in the second approach are *CommonKAD*[25] (which focused on the application of knowledge information systems in a company/organisation), *Methontology*[26], *Web Semantics Design Method* (WSDM)[27], and *Semantic Hypermedia Design Method* (SHDM)[28].

In cultural heritage domain, Freire, Calado, and Martins[29] found that none of URLs, out of 52,866 URLs in Europeana providers, accommodates Semantic Web services with RDF or JSON-LD format. Rather, most of them relies on the format of HTML documents which intended for humans. For data synchronisation, OAI-PMH[e] protocol is widely used,

[e]Open Archives Initiative Protocol for Metadata Harvesting. `http://www.openarchives.org/`

notwithstanding the urge of change OAI-PMH is getting stronger in accordance with the computing performance improvement. The alternatives of aforesaid web data technologies[30] are International Image Interoperability Framework (IIIF), Webmention, Linked Data Notifications, WebSub, Sitemaps, ResourceSync, Open Publication Distribution System (OPDS), Linked Data Platform, and Schema.org.

The open data and open knowledge principles supports the development of information service networks that links information objects from a number of information source institutions. Accordingly, the Semantic Web framework can be developed into a large framework of links between information providers called *Linked Open Data* (LOD). A challenge in developing a cultural object information repository that applies the Semantic Web and LOD is the cultural object information characteristic[31], i.e. multi-format (documents, films, pictures, etc.), multi-topic, multi-language, multi-cultural, multi-target (content for laymen or experts, young or old). As a consequence, vocabulary control of particular cultural object is significant. A taxonomy of Indonesian cultural object proposed by Proboyekti and Susanto in Ref. 32 consists of 17 categories and 80 subcategories.

LOD empowers a semantic portal of cultural heritage service as well as a content model definition for metadata representation, ontology, and rules related to cultural objects. Certain benefits of semantic portal from the end users view point[33] are 1) a global display for heterogeneous and distributed content; 2) automatic content aggregation; 3) semantic search; 4) semantic exploration and recommendations; and 5) other intelligent services, e.g. personalisation, semantic visualisation based on history and location. On the other hand, data providers should also enjoy some advantages from the semantic portal, i.e. 1) distributed content creation; 2) automatic link maintenance; 3) sharing publication services together; 4) semantically enrich each other's content; and 5) reuse of content that has been contained in the portal.

Hyvönen in Ref. 33 provides recommendations for each Semantic Web layer of semantic portal appropriate for cultural heritage service. For *metadata schemas* layer, it is recommended to use metadata that supports the description of each cultural object. *Dublin Core (DC) Metadata Element Set*[f] and *Visual Resource Association's (VRA) Core Categories*[g] are potential alternatives as a basis for more detailed metadata schemes. The

[f]http://dublincore.org/documents/dces/
[g]http://www.vraweb.org/

metadata can also be expressed in the form of Turtle notation[h].

At *ontology* layer, it is necessary to provide standards and instructions for terms selection to complete each metadata element. Standard data values are usually determined from controlled vocabularies and thesauri[34,35], e.g. *Thesaurus for Graphic Materials I*[i] (TGM I), ICONCLASS[j], *Art and Architecture Thesaurus*[k] (AAT), *Union List of Artist Names*[l] (ULAN), UNESCO *Thesaurus*[m], *Thesaurus of Geographic Names*, and *Library of Congress Authority Files*[n]. *Cataloging Cultural Objects*[o] (CCO) guidelines is an example of data content standard for descriptive cataloging of cultural works.

In spite of the fact that thesaurus is significant in improving the performance of retrieval engine as well as indexer, constructing a thesaurus for cultural objects is challenging for it should consider multidimensional views. Some thesaurus have been transformed into SKOS[p] format on the grounds that it supports LOD standardisation. A study[36] of LOD KOS found that : 1) LOD KOS vocabulary encourage the data provided to receive 4-5 LOD stars; 2) the LOD approach leads to unconventional processes and results since vocabulary producers involved in developing and enriching KOS; and 3) in terms of end-users products, LOD is a knowledge base which provides semantic-rich discoveries.

In Semantic Web framework, the definition of each cultural object follows its ontology concept. There are 6 ontology types[31] related to the object collection of cultural heritage information classified by their domain of discourse, i.e. 1) General concept ontologies; 2) Actor ontologies; 3) Place ontologies; 4) Time and period ontologies; 5) Event ontologies; and 6) Domain nomenclatures or terminologies. *CIDOC Conceptual Reference Model* (CRM) is the existence ISO21127-standardised ontology reference for cultural heritage. A study of Ariyani and Yuhana[37] successfully linked CIDOC-CRM metadata with the open linked data *Geonames* dataset in order to enrich the spatial information of cultural objects.

The effort of Indonesian cultural objects ontology development ex-

[h] `http://www.dajobe.org/2004/01/turtle/`
[i] `http://www.loc.gov/rr/print/tgm1/`
[j] `http://www.iconclass.nl/`
[k] `http://www.getty.edu/research/conducting_research/vocabularies/aat/`
[l] `http://www.getty.edu/vow/ULANSearchPage.jsp`
[m] `http://vocabularies.unesco.org/browser/thesaurus/en/`
[n] `http://authorities.loc.gov/`
[o] `http://vraweb.org/resources/cataloging-cultural-objects/`
[p] Simple Knowledge Organization System.

pressed by several studies, e.g. batik object ontology in Ref. 38, medicinal plants ontology in Ref. 39, traditional herbal medicine ontology in Ref. 40. These ontologies have no connection to form linked data. This article shows an ongoing study of interconnected ontologies for Indonesian cultural objects. It consists of 13 ontologies and 2 of the ontologies can be found in Ref. 41 and Ref. 42.

3. A Study of LOD-Based Semantic Web

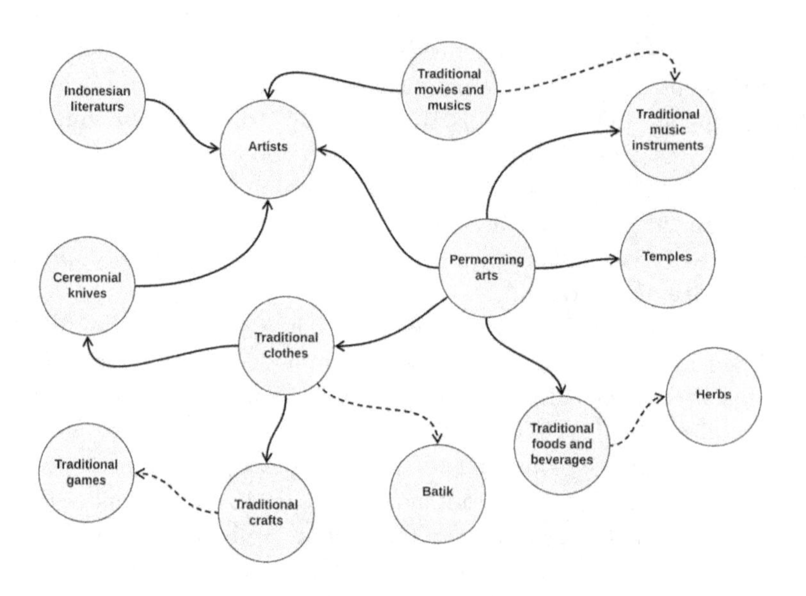

Fig. 1. Ontology modular model of Indonesian cultural objects.

This study was started by focus group discussions (FGDs) consisted of important partners in Special Region of Yogyakarta[q], i.e. Dinas Kebudayaan (Disbud) DIY[r], Badan Perpustakaan dan Arsip Daerah (BPAD)

[q]*Daerah Istimewa Yogyakarta (DIY).*
[r]Department of Culture

DIY[s], Forum Perpustakaan Perguruan Tinggi Indonesia (FPPTI) DIY[t], as well as pertinent practitioners and researchers. The main outcomes of the FGDs were:

(1) Determination of 13 cultural objects selected from Indonesian cultural taxonomy[32]. Those are traditional music instruments (*alat musik tradisional*), temples (*candi*), Indonesian movies and musics (*film dan musik Indonesia*), Indonesian traditional crafts (*kerajinan tradisional Indonesia*), ceremonial knives (*keris*), traditional foods and beverages (*makanan dan minuman tradisional*), traditional clothes (*pakaian adat*), artists (*pelaku seni*), traditional games (*permainan tradisional*), Indonesian literatures (*sastra Indonesia*), performing arts (*seni pertunjukan*), batik (*batik*), and herbs (*jamu*).
(2) A dataset collection of those selected cultural objects.

Table 1. List of IRI ontologies.

No.	Cultural Objects	IRI
1	Traditional music instruments	`http://aunalun.info/ontology/alatmusik`
2	Temples	`http://aunalun.info/ontology/candi`
3	Indonesian movies and musics	`http://aunalun.info/ontology/musikfilm`
4	Traditional crafts	`http://aunalun.info/ontology/kerajinan`
5	Ceremonial knives	`http://aunalun.info/ontology/keris`
6	Traditional foods and beverages	`http://aunalun.info/ontology/makanan`
7	Traditional clothes	`http://aunalun.info/ontology/pakaianadat`
8	Artists	`http://aunalun.info/ontology/seniman`
9	Traditional games	`http://aunalun.info/ontology/permainan`
10	Indonesian literatures	`http://aunalun.info/ontology/sastra`
11	Performing arts	`http://aunalun.info/ontology/senipertunjukan`
12	Batik	`http://aunalun.info/ontology/batik`
13	Herbs	`http://aunalun.info/ontology/jamu`

The ontology modelling is modular in the sense that each ontology of

[s]Regional Library and Archives Agency
[t]Indonesian University Library Forum

cultural object is independent yet having relation each other as it shows in Fig. 1. Each ontology has the same IRI basis, i.e. `http://aunalun.info/ontology` followed by ontology name[u] of cultural object. Table 1 presents the complete IRI ontologies.

The study follows a process described in Fig. 2 which is an adaptation of Methontology and WSDM methodologies. *Knowledge extraction* phase in Fig. 2 is basically a literature study. In this phase, significant sentences were selected or generated by summarizing passages, hence the phase yielded simple sentences comprised of subject, predicate, and object. Table 2 lists all books being extracted for each relevant ontology. Unlike the others, *Traditional foods and beverages* object exploited individual notes taken from observation in traditional markets.

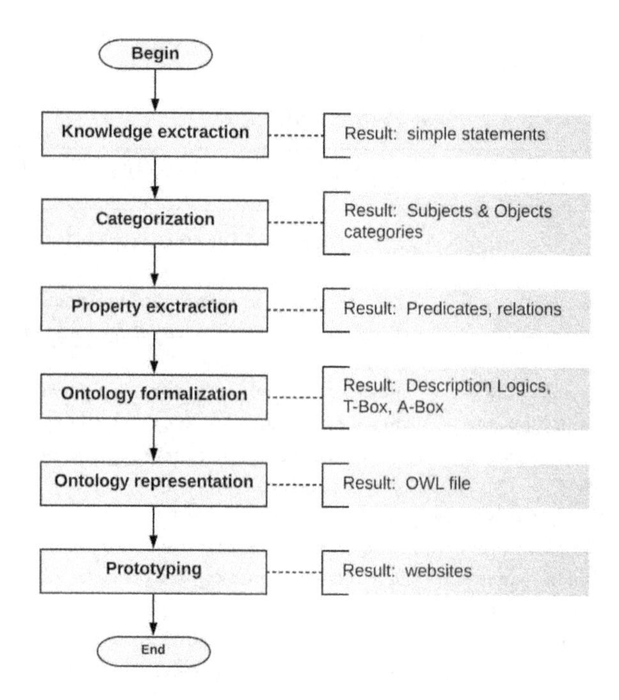

Fig. 2. Research Methodology.

[u]In Indonesian.

Table 2. List of books used as knowledge source.

No.	Cultural Objects	Book Titles
1	Traditional music instruments	*Bentuk-Bentuk Peralatan Hiburan dan Kesenian Tradisional Daerah Istimewa Yogyakarta* [43] *Ensiklopedia Alat Musik Tradisional* [44] *Ensiklopedia Alat Musik Tradisional* [45]
2	Temples	*Indonesian Heritage: Sejarah Awal* [46] *World Heritage Sites and Living Culture of Indonesia* [47]
3	Indonesian movies and musics	*Poster Film Indonesia: Masa Sesudah Kemerdekaan* [48] *Poster Film Indonesia: Masa 1980-1990* [49] *Kumpulan Lagu Daerah Nusantara Terlengkap!* [50]
4	Indonesian traditional crafts	*Album Kerajinan Tradisional (Bengkulu, DKI Jakarta, Jawa Tengah, Jawa Timur, Kalimantan Barat)* [51] *Album Gerabah Tradisional Kasongan Yogyakarta* [52] *Pengrajin Tradisional di Daerah Bali* [53]
5	Ceremonial knives	*Ensiklopedi Keris* [54]
6	Traditional foods and beverages	-
7	Traditional clothes	*Busana Tradisional* [55] *Untaian Manik-manik Nusantara* [56] *Perhiasan Tradisional Indonesia* [57] *Album Pakaian Tradisional Yogyakarta* [58] *Tutup Kepala Tradisional Jawa* [59]
8	Artists	*Seni Rupa* [60]
9	Traditional games	*Permainan Tradisional Indonesia* [61]
10	Indonesian literatures	*Majas, Pantun, dan Puisi* [62]
11	Performing arts	*Indonesia Heritage (Vol. 08)* [63] *Masa Gemilang dan Memudar Wayang Wong Gaya Yogyakarta* [64] *Mengenal Tari-Tarian Rakyat di Daerah Istimewa Yogyakarta* [65]
12	Batik	*Batik Filosofi, Motif dan Kegunaan* [66] *Keeksotisan Batik Jawa Timur: Memahami Motif dan Keunikannya* [67]
13	Herbs	*Empat Puluh Resep Ampuh Tanaman Obat Untuk Mempercepat Kehamilan* [68] *Empat Puluh Resep Dahsyat Jamu Penakluk Asam Urat Dan Diabetes* [69] *Empat Puluh Resep Ampuh Tanaman Obat Untuk Mengobati Jantung Koroner & Menyembuhkan Stroke* [70]

The subjects and objects resulted from previous phase were categorized in order to attain to a list of concepts, i.e. classes or instances, for each cultural objects. Properties of cultural objects, i.e. relations, were derived by summarizing, grouping, as well as categorizing the sentence predicates. After this phase, it was possible to compile a knowledge in the form of description logics and T-box RDFS model for each cultural objects. These were the basis of each ontology which was realized into Protégé. The OWL file yielded by Protégé were functioned as the knowledge in Semantic Web of cultural objects. In our project, GraphDB was used as the SPARQL End Point as it is depicted in Fig. 3.

Fig. 3. GraphDB screenshot of `alunalun.info`.

The following describes the number of classes and triples[v] belongs to each cultural objects:

(1) Traditional music instruments : 16 classes and 4,247 triples
(2) Temples : 45 classes and 5,453 triples
(3) Indonesian movies and musics : 16 classes and 3,884 triples
(4) Traditional crafts : 16 classes and 3,362 triples
(5) Ceremonial knives : 144 classes and 19,219 triples
(6) Traditional foods and beverages : 19 classes and 3,991 triples
(7) Traditional clothes : 16 classes and 2,500 triples
(8) Artists : 25 classes and 3,134 triples

[v] A triple is a set of three elements: a subject, a predicate, and an object.

(9) Traditional games : 18 classes and 9,167 triples
(10) Indonesian literatures : 48 classes and 10,348 triples
(11) Performing arts : 36 classes and 3,600 triples
(12) Batik : 13 classes and 8,126 triples
(13) Herbs : 18 classes and 4,297 triples

4. Portal Applications of Indonesian Cultural Objects

Two ontologies of Indonesian cultural objects have been constructed, i.e. batik and herbs, and implemented on websites. The *Batik Indonesia* website and *Jamu Tradisional Indonesia* are available in `http://alunalun.info/batik/` and `http://alunalun.info/jamu/`, successively. Figure 4 shows the visualization of the batik ontology implemented on its website.

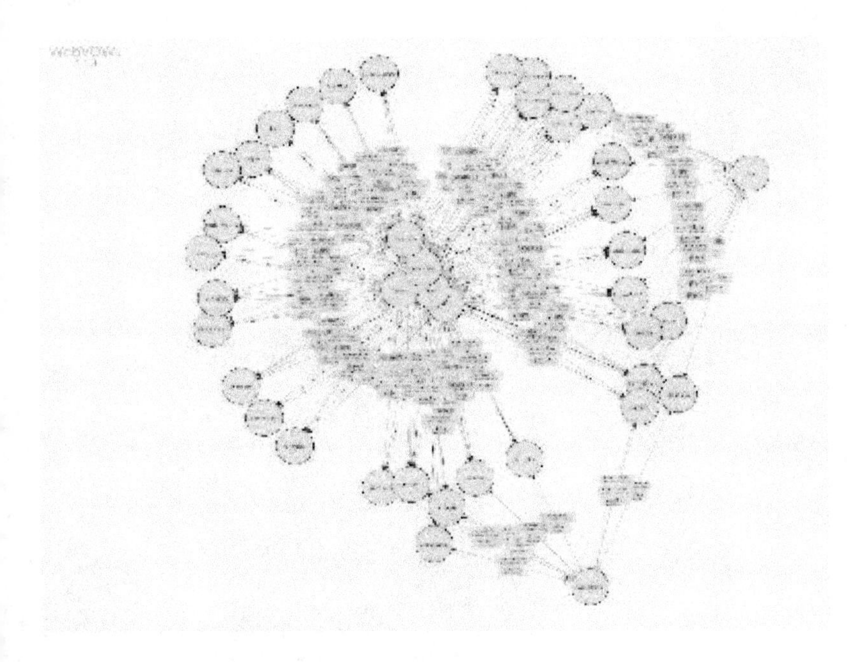

Fig. 4. Visualization of batik ontology.

The 11 other ontologies have been developing and being implemented. The intended portal applications should provide a usable interface for information search and retrieval. Figure 5 to Fig. 8 presents homepage of 4 cultural objects, i.e. ceremonial knives, traditional foods & beverages, Indonesian literatures, and performing arts, successively.

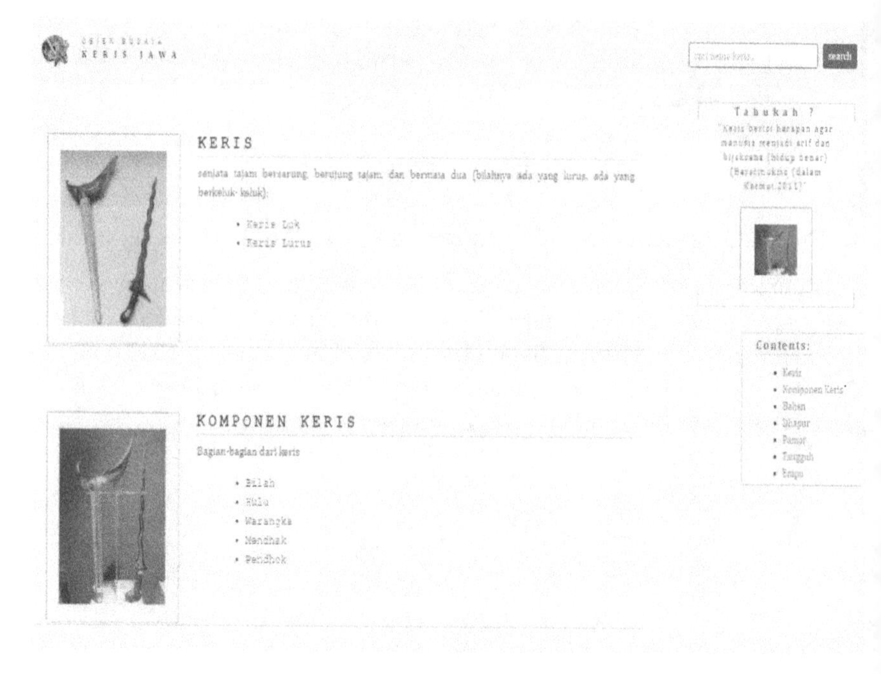

Fig. 5. Homepage application of ceremonial knives object.

Fig. 6. Homepage application of traditional foods & beverages object.

Sastra Indonesia — Home — Sastra Tulis — Sastra Lisan

S a s t r a

INDONESIA

Search

Sastra

Sumber: https://id.wikipedia.org/wiki/Sastra

© 2019 Budi Susanto Gloria Virginia Umi Proboyekti Reiner Sandrico Anglia

Fig. 7. Homepage application of Indonesian literatures object.

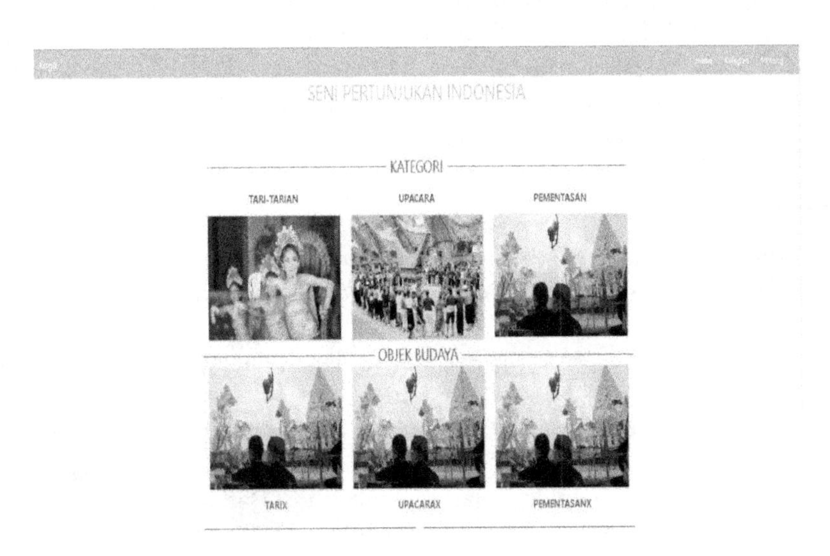

Fig. 8. Homepage application of performing arts object.

5. Conclusion and Future Studies

Various cultural objects, including works from the local community, are local knowledge or local wisdom preserved in the community. Our study is an effort to be an active participant in introducing and disseminating cultural knowledge. It is an endeavour to develop LOD-based Semantic Web for Indonesian cultural objects. Out of 13 modules, 2 ontologies of cultural objects have been successfully constructed, i.e. batik and herbs, while the other ontologies are still in the development process. Nevertheless, homepage of several intended applications presented to show current progress achieved.

The knowledge extraction phase basically was conducted based on selected literatures. In the future, experts of each objects should be involved. The effectiveness and efficiency of the ontologies need to be measured carefully for the quality of the Semantic Web. Another concern is the user interface of the applications in order to ensure users having a good experience.

Acknowledgments

This work is fully supported by the LPPM UKDW under contract number 578/D.01/LPPM/2018 and the Research, Technology, and Higher Education Ministry under contract number 109/SP2H/LT/DRPM/2018. We are grateful for the good collaboration of our partners, i.e. Dinas Kebudayaan DIY, BPAD DIY, and FPPTI DIY.

References

1. Badan Pusat Statistik, *Kewarganegaraan, Suku Bangsa, Agama, dan Bahasa sehari-hari Penduduk Indonesia* (Badan Pusat Statistik, Jakarta, 2011).
2. Pusat Data dan Statistik Pendidikan dan Kebudayaan, Statistik kebudayaan 2016 (2016), Retrieved on October 02, 2018 from `http://publikasi.data.kemdikbud.go.id/uploadDir/isi_5808B5CD-F78A-4A7C-A886-3DB9D1CF688B_.pdf`.
3. Kementerian Pendidikan dan Kebudayaan, Daftar warisan budaya takbenda indonesia tahun 2017 (October 2017), Retrieved on October 02, 2018 from `http://kemdikbud.go.id/main/files/download/6ee499fb7955e5b`.
4. P. T. Juniman, Indonesia tetapkan 150 warisan budaya takbenda (September 2017), Retrieved on October 02, 2018 from `https://www.`

cnnindonesia.com/hiburan/20170927152207-241-244383/indonesia-tetapkan-150-warisan-budaya-takbenda.

5. Kantor Delegasi Tetap Republik Indonesia untuk UNESCO (KWRIU), Warisan budaya tak benda (wbtb) indonesia (September 2017), Retrieved on October 02, 2018 from http://kwriu.kemdikbud.go.id/info-budaya-indonesia/warisan-budaya-tak-benda-indonesia/.
6. D. A. Tanudirjo, Warisan budaya untuk semua: Arah kebijakan pengelola warisan budaya indonesia di masa mendatang (2003), Retrieved on October 02, 2018 from http://arkeologi.fib.ugm.ac.id/old/download/1211776349daud-kongres%20kebud.pdf.
7. Semanticweb.org, Semantic web standards Retrieved on October 03, 2018 from http://semanticweb.org/wiki/Semantic_Web_standards.html.
8. Open data handbook documentation, release 1.0.0 (November 2012), Retrieved on October 26, 2014 from http://opendatahandbook.org/pdf/OpenDataHandbook.pdf.
9. D. Bennett and A. Harvey, Publishing open government data (September 2009), Retrieved on October 26, 2014 from http://www.w3.org/TR/gov-data/.
10. A.-L. Westrum, The key to the future of the library catalog is openness, *Computers in Libraries* **31**, 12 (April 2011).
11. J. Bowen and P. E. Schreur, Linked data for libraries: Why should we care? where should we start? (March 2012), Retrieved on February 20, 2014 from http://www.cni.org/topics/information-access-retrieval/linked-data-for-libraries/.
12. T. Berners-Lee, J. Hendler and O. Lassila, The semantic web., *Scientific American* **284**, 29 (May 2001).
13. M. Sabou, *An Introduction to Semantic Web Technologies*, in *Semantic Web Technologies for Intelligent Engineering Applications*, eds. S. Biffl and M. Sabou (Springer International Publishing, Cham, 2016), Cham, pp. 53–81.
14. V. Sugumaran and J. A. Gulla, *Applied Semantic Web Technologies: Overview and Future Directions*, in *Applied Semantic Web Technologies*, eds. V. Sugumaran and J. A. Gulla (CRC Press, Taylor & Francis Group, Boca Raton, FL, 2012), Boca Raton, FL, pp. 3–17.
15. K. K. Breitman, M. A. Casanova and W. Truszkowski, *Semantic Web: Concepts, Semantic Web: Concepts* (Springer-Verlag, London, 2007).
16. R. Akerkar, *Ontology: Fundamentals and Language*, in *Applied Semantic Web Technologies*, eds. V. Sugumaran and J. A. Gulla (CRC Press,

Taylor & Francis Group, Boca Raton, FL, 2012), Boca Raton, FL, pp. 21–64.

17. A. Gómez-Pérez, M. Fernández-López and O. Corcho, *Ontological Engineering: with Examples from the Areas of Knowledge Management, E-Commerce and the Semantic Web* (Springer-Verlag, London, 2004).
18. S. Grimm, A. Abecker, J. Völker and R. Studer, *Ontologies and the Semantic Web*, in *Handbook of Semantic Web Technologies*, eds. J. Domingue, D. Fensel and J. A. Hendler (Springer-Verlag, Berlin, 2011), Berlin, pp. 509–579.
19. T. R. Gruber, A translation approach to portable ontology specifications, *Knowledge Acquisition* **5**, 199 (1993).
20. F. Baader, I. Horrocks and U. Sattler, *Description Logics*, in *Handbook on Ontologies*, eds. S. Staab and R. Studer (Springer-Verlag, Berlin, 2009), Berlin, pp. 21–43.
21. S. Cranefield and M. Purvis, Uml as an ontology modelling language, in *IJCAI '99 Workshop on Intelligent Information Integration*, eds. D. Fensel, C. Knoblock, N. Kushmeric and M.-C. Rousset (Amsterdam, 1999). http://ceur-ws.org/Vol-23/.
22. D. B. Lenat and R. V. Guha, *Building Large Knowledge-Based Systems; Representation and Inference in the Cyc Project* (Addison-Wesley Longman Publishing Co., Inc., Boston, MA, USA, 1989).
23. M. Grüninger and M. S. Fox, Methodology for the design and evaluation of ontologies, in *IJCAI '95 Workshop on Basic Ontological Issues in Knowledge Sharing*, 1995.
24. G. S. Bob, B. Wielinga and W. Jansweijer, The kactus view on the 'o' word, in *IJCAI '95 Workshop on Basic Ontological Issues in Knowledge Sharing*, 1995.
25. G. Schreiber, H. Akkermans, A. Anjewierden, R. D. Hoog, N. Shadbolt, W. V. d. Velde and B. Wielinga, *Knowledge Engineering and Management: The CommonKADS Methodology* (The MIT Press, Cambridge, Massachusetts, 1999).
26. M. Fernández-Lópe, A. Gómez-Pérez and N. Juristo, Methontology: From ontological art towards ontological engineering, in *Proceedings of the Ontological Engineering AAAI-97 Spring Symposium Series*, March 1997.
27. O. D. Troyer, S. Casteleyn and P. Plessers, *WSDM: Web Semantics Design Method*, in *Web Engineering: Modelling and Implementing Web Applications*, eds. G. Rossi, O. Pastor, D. Schwabe and L. Olsina (Springer-Verlag, London, 2008), London, pp. 303–351.

28. D. Schwabe, G. Szundy, S. S. de Moura and F. Lima, Design and implementation of semantic web applications, in *WWW Workshop on Application Design, Development and Implementation Issues in the Semantic Web*, eds. C. Bussler, S. Decker, D. Schwabe and O. PastorMay 2004. http://ceur-ws.org/Vol-105/.
29. N. Freire, P. Calado and B. Martins, Availability of cultural heritage structured metadata in the world wide web, in *International Conference on Electronic Publishing (ElPub) 2018*, June 2018. https://elpub.episciences.org/4608.
30. F. Nuno, A. Isaac, G. Robson, J. B. Howard and H. Manguinhas, A survey of web technology for metadata aggregation in cultural heritage, *Information Services & Use* **37**, 425 (2017).
31. E. Hyvönen, *Publishing and Using Cultural Heritage Linked Data on the Semantic Web* (Morgan & Claypool, Palo Alto, CA, USA, October 2012).
32. U. Proboyekti and B. Susanto, Kajian kategori objek budaya yogyakarta untuk yogyasiana (2014), Unpublished.
33. E. Hyvönen, *Semantic Portals for Cultural Heritage*, in *Handbook on Ontologies*, eds. S. Staab and R. Studer (Springer Berlin Heidelberg, Berlin, Heidelberg, 2009), Berlin, Heidelberg, pp. 757–778.
34. A. Gilchrist, D. Bawden and J. Aitchison, *Publishing and Using Cultural Heritage Linked Data on the Semantic Web* (Aslib IMI, The Association for Information Management, London, 2005).
35. R. Baeza-Yates and B. Ribeiro-Neto, *Modern Information Retrieval*, 2nd edn. (Addison-Wesley Professional, 2011).
36. M. L. Zeng and P. Mayr, Knowledge organization systems (kos) in the semantic web: A multi-dimensional review, *International Journal on Digital Libraries* **20**, 209 (2019).
37. N. F. Ariyani and U. L. Yuhana, Generating cultural heritage metadata as linked open data, in *International Conference on Information Technology Systems and Innovation (ICITSI)*, November 2015.
38. T. K. Rahayu, E. Sediyono and O. D. Nurhayati, Semantic search based on ontology with case study: Indonesian batik, *International Journal of Computer Applications* **170**, 20 (July 2017).
39. D. W. Wardani, S. H. Yustianti, U. Salamah and O. P. Astirin, An ontology of indonesian ethnomedicine, in *International Conference on Information Communication Technology and System*, September 2014.
40. R. Gunawan and K. Mustofa, Semantic search based on ontology with case study: Indonesian batik, *International Journal of Electrical and*

Computer Engineering **7**, 3674 (December 2017).

41. B. Susanto, H. Y. Situmorang and G. Virginia, Rdfs-based information representation of indonesian traditional jamu, in *Lecture Notes in Engineering and Computer Science: Proceedings of The International MultiConference of Engineers and Computer Scientists 2019, IMECS 2019*, 13-15 March 2019.
42. B. Susanto, B. Valgian, G. Virginia and U. Proboyekti, Perancangan awal model pengetahuan batik indonesia berbasis semantic web, in *Prosiding Seminar Nasional Teknologi Informasi dan Komunikasi (SEMNASTIK) X*, October 2018.
43. Moertjipto, Y. Ahmad and S. Hp., *Bentuk-Bentuk Peralatan Hiburan dan Kesenian Tradisional Daerah Istimewa Yogyakarta* (Departemen Pendidikan dan Kebudayaan, Direktorat Jenderal Kebudayaan, Direktorat Sejarah dan Nilai Tradisional, Proyek Inventarisasi dan Pembinaan Nilai-Nilai Budaya, 1990).
44. T. Utama, *Ensiklopedia alat musik tradisional* (CV Angkasa, Bandung, 2014).
45. T. O. M, *Ensiklopedia alat musik tradisional* (SIC, Surabaya, 2008).
46. J. Miksic, *Indonesian Heritage: Sejarah Awal* (Buku Antar Bangsa, Jakarta, 2002).
47. *World Heritage Sites and Living Culture of Indonesia* (Ministry of Education and Culture, Republic of Indonesia, Jakarta, 2012).
48. A. Pranajaya, W. T. Haryanti and D. Isyanti, *Poster Film Indonesia: Masa Sesudah Kemerdekaan*, in *Poster Film Indonesia: Masa Sesudah Kemerdekaan*, eds. A. Pranajaya, W. T. Haryanti and D. Isyanti (Perpustakaan Nasional Republik Indonesia, Jakarta, 2010), Jakarta, pp. 76–76.
49. A. Pranajaya, W. T. Haryanti and D. Isyanti, *Poster Film Indonesia: Masa 1980-1990*, in *Poster Film Indonesia: Masa Sesudah Kemerdekaan*, eds. A. Pranajaya, W. T. Haryanti and D. Isyanti (Perpustakaan Nasional Republik Indonesia, Jakarta, 2011), Jakarta, pp. 74–74.
50. T. Khaira (ed.), in *Kumpulan Lagu Daerah Nusantara Terlengkap!*, ed. T. Khaira (Dar! Mizan, Bandung, 2009), Bandung.
51. Proyek Pengembangan Media Kebudayaan (Indonesia), *Album Kerajinan Tradisional (Bengkulu, DKI Jakarta, Jawa Tengah, Jawa Timur, Kalimantan Barat)* (Direktorat Jenderal Kebudayaan, Jakarta, 1993).
52. S. Wisetrotomo, *Album Gerabah Tradisional Kasongan Yogyakarta* (Direktorat Jenderal Kebudayaan, Jakarta, 1995).

53. I. N. Martha, *Pengrajin Tradisional di Daerah Bali* (Direktorat Jenderal Kebudayaan, Direktorat Sejarah dan Nilai Tradisional, Jakarta, 1991).
54. B. Harsrinuksmo, *Ensiklopedi Keris* (Gramedia Pustaka Utama, Jakarta, 2004).
55. Yayasan Harapan Kita, *Busana Tradisional* (Yayasan Harapan Kita, Jakarta, 1998).
56. Hamzuri and T. R. Siregar, *Untaian Manik-manik Nusantara* (Direktorat Jenderal Kebudayaan, Jakarta, 1999).
57. M. Husni and T. R. Siregar, *Perhiasan Tradisional Indonesia* (Direktorat Jenderal Kebudayaan, Jakarta, 2000).
58. A. M. Hidayati, *Album Pakaian Tradisional Yogyakarta* (Direktorat Jenderal Kebudayaan, Jakarta, 1992).
59. S. Tukio, *Tutup Kepala Tradisional Jawa* (Direktorat Jenderal Kebudayaan, Jakarta, 1981).
60. H. Soemantri, W. P. Wibowo, K. H. Saputra and Y. Affendi, *Seni Rupa* (Buku Antar Bangsa, Jakarta, 2002).
61. Hamzuri and T. R. Siregar, *Permainan Tradisional Indonesia* (Direktorat Jenderal Kebudayaan, Jakarta, 1998).
62. U. N. Masruchin, *Majas, Pantun, dan Puisi* (Huta Publisher, Depok, 2017).
63. E. Sedyawati, *Indonesia Heritage (Vol. 8)* (Buku antar Bangsa, Jakarta, 2002).
64. R. M. Soedarsono, *Masa Gemilang dan Memudar Wayang Wong Gaya Yogyakarta* (Tarawang, Yogyakarta, 2000).
65. Soedarsono, *Mengenal Tari-Tarian Rakyat di Daerah Istimewa Yogyakarta* (Akademi Seni Tari Indonesia, Yogyakarta, 1976).
66. A. Kusrianto, *Batik Filosofi, Motif dan Kegunaan* (Penerbit Andi, Yogyakarta, 2014).
67. Y. Anshori and A. Kusrianto, *Keeksotisan Batik Jawa Timur: Memahami Motif dan Keunikannya* (Elex Media Komputindo, Jakarta, 2011).
68. I. Manganti, *Empat Puluh Resep Ampuh Tanaman Obat Untuk Mempercepat Kehamilan* (Pustaka Araska Media Utama, Yogyakarta, 2015).
69. E. Winasis, *Empat Puluh Resep Dahsyat Jamu Penakluk Asam Urat Dan Diabetes* (Pustaka Araska Media Utama, Yogyakarta, 2014).
70. I. Manganti, *Empat Puluh Resep Ampuh Tanaman Obat Untuk Mengobati Jantung Koroner & Menyembuhkan Stroke* (Pustaka Araska Media Utama, Yogyakarta, 2015).

Age Difference of Self Learning Aspect, Effective Learning Aspect and Multimedia Aspect Towards E-Learning Courses in Hong Kong Higher Education

H.K. Yau[†] and K.C. Yeung
City University of Hong Kong,
Kowloon Tong, Kowloon, Hong Kong.
[†]E-mail: honkyau@cityu.edu.hk

The purpose of this study is to examine the differences between ages on Self Learning Aspect (SLA), Effective Learning Aspect (ELA) and Multimedia Aspect (MMA) towards e-learning courses in Hong Kong higher education. 140 questionnaires were collected. The findings show that there are no significantly difference between ages on Self-learning Aspect (SLA), Effective Learning Aspect (ELA) and Multimedia Aspect (MMA) towards e-learning courses in Hong Kong higher education

Keywords: Effective Learning Aspect; Multimedia Aspect; Self Learning Aspect.

1. Introduction

Nowadays, E-learning has become more popular and universal around the world. E-learning is learning utilizing electronic technologies to access educational curriculum outside of a traditional classroom. In most cases, it refers to a course, program or degree delivered. However, all the educations via electronic applications, processes can be defined as E-learning. For instance, Web-based learning, computer-based learning, etc. Although E-learning is mainly focusing on distance education, it is also transforming instruction on campus. Normally, students need to take classes in campus. However, E-learning has blurred these traditional relationships and solved the geographical problems. There are several major goals of E-learning include: improving access for people with different ages, physically disabled to attend traditional courses; increasing student's choices on how, when, where to engage the learning process; and improving the effectiveness and efficiency by using E-learning media and methods. In Hong Kong, E-learning has been used in higher education first for a few years. Nowadays, all the higher education institutions can provide E-learning resources to students as well as teachers. Yau and Yeung [1] have examined **is** to examine the differences between males and females on Self Learning Aspect (SLA), Effective Learning Aspect (ELA) and Multimedia Aspect (MMA) towards e-learning courses in Hong Kong higher education. However, no studies have examined the age difference related to Self Learning Aspect (SLA), Effective Learning Aspect (ELA) and Multimedia Aspect (MMA) towards e-learning courses in Hong Kong higher education. The objective of this paper is to fill this research gap.

2. Literature Review

Students' perceived satisfaction, behavior intention and effectiveness of e-learning may be affected by several variables. For instance, learner's attitudes. Liaw [3] has stated that learner's attitudes are one of the major factors that affect the effectiveness and efficiency of E-learning [3]. Learning's attitudes include age, gender, nationality, level of education, Self- learning aspect (SLA), Effective learning aspect (ELA) and Multimedia aspect (MMA).

The independent variables, such as gender, age, nationality may affect the students' perceived satisfaction, effectiveness of e-learning. For instance, sex difference will affect people's ways of thinking [2], which may influence the effectiveness of e-learning in this study. People in different ages may also affect their ways of thinking. Older people have relatively low acceptance on learning new things [5]. And also for nationality, Culture affects the learners' behaviors of online education and these effects must be taken into consideration to make e-learning more effective and practical [6].

The main features of the learning process include interactions among students, and interactions between students and instructors for collaboration in learning [4]. In this study, totally 3 hypotheses were made, which are related to age and 3 aspects – Self-learning aspect (SLA), Effective learning aspect (ELA) and Multimedia aspect (MMA) in questionnaire.

Older people have relatively slower perceptual learning than younger ones [7]. Research suggests that age is an important demographic variable that has direct and moderating effects on behavioral intention, adoption, and acceptance of technology [8]. Age difference probably will be a huge factor that affects the effectiveness of E-learning.

In terms of computer and Internet self-efficacy, it has been found that older people have low self-efficacy in use of technology [8], which means in this research, older people may provide a lower rating on e-learning questionnaire as they may have low self- efficacy in use of technology.

Moreover, Tarhini, Hone, & Liu [8] state that age differences influence the perceived difficulty of learning a new software application, and there is clear evidence that younger adults have lower levels of computer anxiety than their older counterparts.

Furthermore, Tarhini, Hone, & Liu [4] also state that there was a moderating effect of age on the relationship between social influence and behavioral intention, with the relationship stronger for older users, which means age difference will probably affect students' behavioral intention on e-learning.

As mentioned above, age difference may cause people's acceptance of e-learning become totally different. Therefore, we hypothesized:

H1: Older students rating of Self-learning aspect (SLA) will be lower than young students

H2: Older students rating of Effective learning aspect (ELA) will be lower than young students

H3: Older students rating of Multimedia aspect (MMA) will be lower than young students

3. Research Methodology

In this study, 140 questionnaires were distributed to students to collect their views on e- learning.. There are 14 questions (Table 1) in total, including: 5 questions in Self- learning aspects (SL1 -5), 5 questions in Effective learning aspects (EL1-5) and 4 questions in Multimedia aspects (MM1-4).

Table 1 – Items in questionnaire

Item	Questions
SL1	E-learning assists my learning more than lessons
SL2	I have more opportunities to learn more by e-learning
SL3	I can find academic materials by e-learning
SL4	Using e-learning and technology can enhance my learning motivation
SL5	I can discuss with foreign people in the e- learning environment
EL1	E-learning can increase my learning efficiency
EL2	E-learning can improve my thinking skill
EL3	E-learning can enhance my problem- solving skills

EL4	E-learning can help to develop my ability to act or think independently
EL5	E-learning provides diversified environment for effective learning
MM1	I like learning video in my studies
MM2	I like learning animation in my studies
MM3	I like using internet to discuss with others
MM4	E-learning is more convenient than traditional learning

For self-learning aspect, the factor loading of SL1 to SL5 are 0.867, 0.916, 0.918, 0.895 and 0.886 respectively.

For effective learning aspect, the factor loading of EL1 to EL5 are 0.895, 0.915, 0.909, 0.913 and 0.872 respectively.

For multimedia aspect, the factor loading of MM1 to MM54 are 0.939, 0.942, .0941 and 0.951 respectively

The above items are retained as all of them are over 0.3 [10][11].

The Cronbach's Alpha values of self-learning aspect, effective learning aspect and multimedia aspect are 0.938, 0.942 and 0.958 respectively. Because all three reliability coefficients are greater than 0.7 which is considered "acceptable" in most social science research situations [9].

4. Results and Discussion

There are total 140 questionnaires collected in this study. About 48.6% of interviewee are male while about 51.4% are female. 6.4% of them are with age <18, 22.9% with age 18, 21.4% with age 19, 20.7% with age 20, 17.9% with age 21, 5% with age 22 and the remaining 5.7% with age >22. Moreover, 45% of interviewees are local students, 30% are from mainland China, 4.3% are from other Asia countries, 5% are from Europe, 6.4% are from America, 5% are from Africa, and the remaining 4.3% are from Oceania. For their level of education, 16.4% of interviewees are Associate degree, 2.1% are Top-up degree, 74.3% are Bachelor degree, 5% are Master degree, and the remaining 2.1% are PhD degree. For their college of study, 22.9 of them are from College of Business, 21.4% of them of from College of Liberal Arts and Social Sciences, 35% of them are of College of Engineering and Sciences, 8.6% of them are from School of Creative Media, 7.1% of them are from School of Energy and Environment, and the remaining 5% are from School of Law. For their academic year, 42.9%

of them are Year 1 students, 18.6% of them are Year 2 students, 17.9% of them are Year 3 students, and the remaining 20.7% are Year 4 students.

In the following section, description statistics of 3 aspects, Self-learning aspect (SLA), Effective learning aspect (ELA) and Multimedia aspect (MMA) will be shown. The mean, standard deviation and variance of every questions in above aspects will be shown in the following sections.

Moreover, the trend that participants answer in different aspects will be shown in the tables also. Through those descriptive statistics, we can observe whether students tend to accept e- learning or not as well as whether e-learning increase their study efficiency.

Table 2: Description Statistics of Self-learning aspect (SLA)

	N	Mean	Std. Deviation	Variance
Self-learning aspect	140	4.6729	1.38693	1.924
Valid N (listwise)	140			

Table 3: Description Statistics of each item in Self-learning aspect (SLA)

	N	Mean	Std. Deviation	Variance
SL1	140	4.5214	1.45665	2.122
SL2	140	4.7143	1.49957	2.249
SL3	140	4.9143	1.59354	2.539
SL4	140	4.6929	1.48806	2.214
SL5	140	4.5214	1.69412	2.870
Valid N (listwise)	140			

From Table 2, the overall mean of 5 questions in Self-learning aspect (SLA) is 4.6729, which is between option "Neither agree nor disagree" and "Somewhat agree", means that interviewees tend to agree E-learning has motivated them in self-learning.

From Table 3, SL3 "I can find academic materials by e-learning" get the highest mean rating with 4.9143 among Self-learning aspect (SLA), which can also prove E-learning has positive impact on self-learning. Although SL1 "E-learning assists my learning more than lessons" and SL5 "I can discuss with foreign people in the e-learning environment" get the lowest mean rating with 4.5214, it still within the option "Neither agree nor disagree" and "Somewhat agree".

Table 4: Description Statistics of Effective learning aspect (ELA)

	N	Mean	Std. Deviation	Variance
Effective learning aspect	140	4.7557	1.42880	2.041
Valid N (listwise)	140			

Table 5: Description Statistics of each item in Effective learning aspect (ELA)

	N	Mean	Std. Deviation	Variance
EL1	140	4.7929	1.63371	2.669
EL2	140	4.5857	1.57309	2.475
EL3	140	4.7357	1.50113	2.253
EL4	140	4.6357	1.60579	2.579
EL5	140	5.0286	1.61799	2.618
Valid N (listwise)	140			

From Table 4, the overall mean of 5 questions in Effective learning aspect (ELA) is 4.7557, which is between option "Neither agree nor disagree" and "Somewhat agree", means that interviewees tend to agree E-learning is effective for their studies.

From Table 5, EL5 "E-learning provides diversified environment for effective learning" get the highest mean rating with 5.0286 among Effective learning aspect (ELA), which can also prove E-learning tend to be effective for students. Although EL2 "E-learning can improve my thinking skills" get the lowest mean rating with 4.5857, it still within the option "Neither agree nor disagree" and "Somewhat agree".

Table 6: Description Statistics of Multimedia aspect (MMA)

	N	Mean	Std. Deviation	Variance
Multimedia aspect	140	5.2357	1.57395	2.477
Valid N (listwise)	140			

Table 7: Description Statistics of each item in Multimedia aspect (MMA)

	N	Mean	Std. Deviation	Variance
MM1	140	5.1643	1.63421	2.671
MM2	140	5.2571	1.64195	2.696
MM3	140	5.2143	1.69547	2.875
MM4	140	5.3071	1.70440	2.905
Valid N (listwise)	140			

From Table 6, the overall mean of 4 questions in Multimedia aspect (MMA) is 5.2357, which is between option "Somewhat agree" and "Agree", means that interviewees agree E- learning with multimedia tools is more effective.

From Table 7, MM34 "E-learning is more convenient than traditional learning" get the highest mean rating with 5.3071 among Multimedia aspect (MMA), which can also prove multimedia is useful for enhancing efficiency of E-learning. Although MM1 "I like learning video in my studies" get the lowest mean rating with 5.1643, it still within the option "Somewhat agree" and "Agree".

Table 8: ANOVA for Age Groups on Self-learning aspect (SLA)

	Sum of Squares	df	Mean Square	F	Sig.
Between Groups	22.787	6	3.798	2.065	.061
Within Groups	244.590	133	1.839		
Total	267.377	139			

Table 9: Decriptives of Age Groups on Self-learning aspect (SLA)

	N	Mean	Std. Deviation	Std. Error	95% Confidence Interval for Mean	
					Lower Bound	Upper Bound
17 or below	9	4.7333	1.32288	.44096	3.7165	5.7502
18.00	32	4.7125	1.39116	.24592	4.2109	5.2141
19.00	30	4.8267	1.21908	.22257	4.3715	5.2819
20.00	29	4.7655	1.43135	.26580	4.2211	5.3100
21.00	25	4.9280	1.26607	.25321	4.4054	5.4506
22.00	7	4.2000	1.79258	.67753	2.5421	5.8579
23 or above	8	3.1500	1.32988	.47018	2.0382	4.2618
Total	140	4.6729	1.38693	.11722	4.4411	4.9046

From Table 8, it shows that the significance level under ANOVA is 0.061 ($F=2.065$, $p>0.05$), which is higher than 0.05, means there are no significant difference between Age and Self-learning aspect (SLA).

From Table 9, it shows the mean value for all the age groups, which are 4.7333 for age 17 or below, 4.7125 for age 18, 4.8627 for age 19, 4.7655 for age 20, 4.9280 for age 21, 4.2 for age 22 and 3.15 for 22 or above. In this result, there are no significant differences we can observed through the increasing age.

According to the results above, the hypothesis **H1: Older students rating of Self-learning aspect (SLA) will be lower than young students**, is rejected.

Table 10: ANOVA for Age Groups on Effective learning aspect (ELA)

	Sum of Squares	df	Mean Square	F	Sig.
Between Groups	16.969	6	2.828	1.410	.215
Within Groups	266.796	133	2.006		
Total	283.765	139			

Table 11: Descriptives of Age Groups on Effective learning aspect (ELA)

	N	Mean	Std. Deviation	Std. Error	95% Confidence Interval for Mean	
					Lower Bound	Upper Bound
17 or below	9	5.1556	1.03333	.34444	4.3613	5.9498
18.00	32	4.8438	1.29289	.22855	4.3776	5.3099
19.00	30	4.8200	1.37098	.25031	4.3081	5.3319
20.00	29	4.7379	1.57308	.29211	4.1396	5.3363
21.00	25	4.9440	1.38145	.27629	4.3738	5.5142
22.00	7	4.4571	1.87159	.70740	2.7262	6.1881
23 or above	8	3.4500	1.50333	.53151	2.1932	4.7068
Total	140	4.7557	1.42880	.12076	4.5170	4.9945

From Table 10, it shows that the significance level under ANOVA is 0.215 (F=1.41, p>0.05), which is higher than 0.05, means there are no significant difference between Age and Effective learning aspect (ELA).

From Table 11, it shows the mean value for all the age groups, which are 5.1556 for age 17 or below, 4.8438 for age 18, 4.82 for age 19, 4.7379 for age 20, 4.944 for age 21, 4.4571 for age 22 and 3.45 for 22 or above. In this result, there are no significant differences we can observed through the increasing age.

According to the results above, the hypothesis **H2: Older students rating of Effective learning aspect (ELA) will be lower than young students**, is rejected.

Table 12: ANOVA for Age Groups on Multimedia aspect (MMA)

	Sum of Squares	df	Mean Square	F	Sig.
Between Groups	25.449	6	4.242	1.769	.110
Within Groups	318.897	133	2.398		
Total	344.346	139			

Table 13: Descriptives of Age Groups on Multimedia aspect (MMA)

	N	Mean	Std. Deviation	Std. Error	95% Confidence Interval for Mean	
					Lower Bound	Upper Bound
17 or below	9	5.7222	1.12808	.37603	4.8551	6.5893
18.00	32	5.2969	1.52590	.26974	4.7467	5.8470
19.00	30	5.4667	1.49386	.27274	4.9089	6.0245
20.00	29	5.0172	1.62147	.30110	4.4005	5.6340
21.00	25	5.4900	1.40772	.28154	4.9089	6.0711
22.00	7	5.0714	1.95104	.73742	3.2670	6.8758
23 or above	8	3.7188	1.99301	.70464	2.0525	5.3850
Total	140	5.2357	1.57395	.13302	4.9727	5.4987

From Table 12, it shows that the significance level under ANOVA is 0.11 ($F=1.769$, $p>0.05$), which is higher than 0.05, means there are no significant difference between Age and Multimedia aspect (MMA).

From Table 13, it shows the mean value for all the age groups, which are 5.7222 for age 17 or below, 5.2969 for age 18, 5.4667 for age 19, 5.0172 for age 20, 5.49 for age 21, 5.0714 for age 22 and 3.7188 for 22 or above. In this result, there are no significant differences we can observed through the increasing age.

According to the results above, the hypothesis **H3: Older students rating of Multimedia aspect (MMA) will be lower than young students**, is rejected.

In this study, there are 7 age groups in the questionnaires which are 17 or below, 18, 19, 20, 21, 22, 23 or above. There are some studies before stated that age difference may affect people's acceptance in e-learning. Tarhini, Hone, & Liu [8] stated that age differences influence the perceived difficulty of learning a new software application, and there is clear evidence that younger adults have lower levels of computer anxiety than their older counterparts. Based on those studies, I hypothesized **H1: Older students rating of Self- learning aspect (SLA) will be lower than young students, H2: Older students rating of Effective learning aspect (ELA) will be lower than young students, H3: Older students rating of Multimedia aspect (MMA) will be lower than young students.**

However, the result find that neither H1, H2 nor H3 are rejected, which means this study shows that age difference will not affect their rating on both Self-learning aspect (SLA), Effective learning aspect (ELA) and Multimedia aspect (MMA). There are several reasons that may cause this phenomenon. At first, the sample size of the questionnaire distribution is too small when compare to previous studies. In this study, only 140 questionnaires were distributed, which will affect the result.

Moreover, the range of age may be too small in this study. As the major respondents in this study is students, the data collected in range of age are mostly within 17-23 years old.

Compare with previous studies, most of them provided a wider age range for respondents to choose. For instance, Islam, Rahim, Liang, & Momtaz [7]'s study provided a wider age range for respondents such as 30 and above as their targets are not only students.

5. Conclusions

It can be concluded that the three hypotheses in this paper are all rejected. Those information above is useful for higher education institutions in Hong Kong as well as Hong Kong Government to enhance or facilitate the development of e-learning in Hong Kong.

The limitations of this study are that only 140 questionnaires were distributed to students who are studying Bachelor degree, which cannot cover students with all level of educations. If sufficient resource is provided, the sample size is enlarged which the education level will be more evenly distributed.

Apart from the factors of age, future studies can be conducted to investigate other different factors, for instance, such as study mode and nationality of students as well as academic performances.

References

1. H. K. Yau, and K. C. Leung, Gender Difference of self learning aspect, effective learning aspect and multimedia aspect towards e-learning courses in Hong Kong higher education. Lecture Notes in Engineering and Computer Science Proceedings of the International MultiConference of Engineers and Computer Scientists 2019, IMECS 2019, 13-15 March, Hong Kong, pp. 408-410.

2. G. Hofstede, *Culture's Consequences: International Differences in Work-Related Values. (*SAGE, 1984).

3. S. S. Liaw, Considerations for Developing Constructivist Web-Based Learning. *International Journal of Instructional Media*, (2004) pp. 309-321.

4. R.M. Palloff, & K. Pratt, *Building Online Learning Communities EFFECTIVE STRATEGIES FOR THE VIRTUAL CLASSROOM. (*San Francisco, Jossey-Bass, 1999).

5. B. Khan, *Managing E-learning: Design, Delivery, Implementation, and Evaluation. (*Idea Group Inc (IGI), 2005).

6. N. HAMEED, M. U. SHAIKH, F. HAMEED, & A. SHAMIM, *Cornell University Library. (21* March 2016). .Retrieved from Cornell University : https://arxiv.org/pdf/1607.01359.pdf

7. M.A. Islam, N.A.A. Rahim, T.C. Liang, & H. Momtaz, Effect of Demographic Factors on E-learning Effectiveness in a Higher Learning Institution in Malaysia, *International Education Studies,* **4**, 1, (2011), pp. 112-121.

8. Tarhini, A., Hone, K., & Liu, X, Measuring the Moderating Effect of Gender and Age on E-learning Acceptance in England: A Structural equation Modeling Approach for an Extended Technology Acceptance Model, *Journal of Educational Computing Research,* **51**, 2, (2014), pp. 163-184.

9. UCLA Institute for Digital Research and Education, *WHAT DOES CRONBACH'S ALPHA MEAN? | SPSS FAQ* . Retrieved from UCLA Institute for Digital Research and Education: (2017) http://stats.idre.ucla.edu/spss/faq/what-does-cronbachs-alpha-mean/

10. F.J. Fowler, *Survey Research Methods 4th edition*, (US, SAGE Publications Inc., 2009)

11. R. Sapsford, *Survey Research (2nd. Edition),* (London, SAGE Publications, 2007).

Design the Process Separation of Methylal/Methanol by Response Surface Methodology Based on Process Simulation in Extractive Distillation*

W. Weerachaipichasgul†, A. Wanwongka, S. Saengdaw, and A. Chanpirak
Division of Chemical Engineering, Department of Industrial Engineering, Faculty of Engineering, Naresuan University, Phitsanulok, 65000 Thailand,
†E-mail: weerawunw@nu.ac.th (Corresponding)
www.nu.ac.th

P. Kittisupakorn
Department of Chemical Engineering, Faculty of Engineering, Chulalongkorn University, Bangkok, 10330, Thailand
E-mail: Paisan.K@chula.ac.th

Methylal can be blended in diesel oil to reduce gas pollutants and to improve performance of a diesel engine. Methylal (Dimethoxymethane, DMM) can be synthesized by the reversible reaction of methanol excess and formaldehyde or paraformaldehyde to eliminate the chemical equilibrium. A mixture of methylal and methanol can be separated by extractive distillation system that combines with an extractive distillation (EDC) and entrainer recovery (RDC) column. Disadvantage of the extractive distillation system is the great energy consumption especially in the EDC. The design of EDC needs to be optimized the energy consumption and product purity. In this work, a conventional entrianer (DMF) was substituted by a lower-toxic of entrainer such as propylene-glycol. The response surface methodology (RSM) was used to optimize the condition in EDC design. The parameters and their interactions had significant effect on energy consumption in the reboiler and product purity. The optimal parameters of EDC could be determined by RSM with simulation runs and the predicted results by the RSM gave good agreement with the simulated results. Purity of 99.90 % methylal was withdrawn in the overhead of the EDC that could be achieve.

Keywords: Azeotropic mixture; Extractive distillation; Propylene glycol; Response surface methodology

* This work is supported by the Research Council of Naresuan University (R2562E023).

1. Introduction

Distillation is a common process for a thermal separation technology which is based on the different boiling point that is widely applied in chemical manufacturers. However, distillation process needs highly energy consumption and the azeotropic mixture cannot be isolated by conventional distillation column. Although there are many separation techniques for the azeotropic mixture such as membrane pervaporation, pressure swing distillation, azeotropic and extractive distillation, an extractive distillation is the most frequently used because of a greater variety of entrainers and a wider range of operation conditions [4].

Methylal (Dimethoxymethane, DMM) is an intermediate agent that consists of 42 % wt. oxygen, 100% miscible with diesel, and low soot in the combustion. Therefore, methylal can be used to be additive in diesel fuel to reduced gas pollutants [5-6]. Methylal can be produced by reversible reaction with catalyst of methanol and formaldehyde or paraformaldehyde. Since the yield of methylal is limited chemical equilibrium, an excess of methanol is fed to the reactor for killing the chemical equilibrium [7-9]. Therefore, the mixture of methylal and methanol will be separated by an extractive distillation system because the type of mixture is a minimum-boiling azeotrope at atmosphere pressure with 94.06 wt. % methylal. Normally, an entrainer selection as low toxicity, easy recovery, thermal stability, high boiling point, high relative volatility between key components, and high capacity will be considered. In the extractive distillation system of mixture of methanol and methylal, dimethylformamide (DMF), the mixture of DMF and ionic liquid, and ethylene glycol [10 - 12] have been selected to be the entrainer. Although the selectivity of methylal to methanol of DMF is higher than that ethylene glycol, the ethylene glycol is the lower toxicity than that DMF [11]. Tetra-ethylene-glycol can be selected as entrainer but energy consumption is higher than that ethylene glycol usage [13]. Because the boiling point of tetra-ethylene-glycol is higher than that ethylene glycol. The properties of propylene glycol are like ethylene glycol and the toxicity of propylene glycol is lower than that ethylene glycol.

Therefore, the purpose of this work is the design an extractive distillation process to separate methylal by using propylene glycol. Therefore, the vapor-liquid equilibrium (VLE) for the methanol-methylal-propylene glycol system at 1 atm needs to study and compares with ethylene glycol. Moreover, response surface methodology will be applied to find the optimal structural parameters and operational conditions in terms of product purity and energy consumption.

2. Steady State Design

The process flowsheet of the extractive distillation as showed in Fig. 1 that has two columns, one for extractive distillation column (EDC) and another for entrainer recovery column (RDC). The azeotropic mixture (F) and the entrainer (E) streams are fed to extractive distillation column (EDC), where the desired compound takes place to the top of the extractive distillation column (D1). The bottom product of the extractive distillation column (B1) feeds to the RDC, where the entrainer is recycled to the EDC. However, some entrainer losses in the D1 and D2 streams, as a result, the makeup stream of entrainer needs to balance in the system [13].

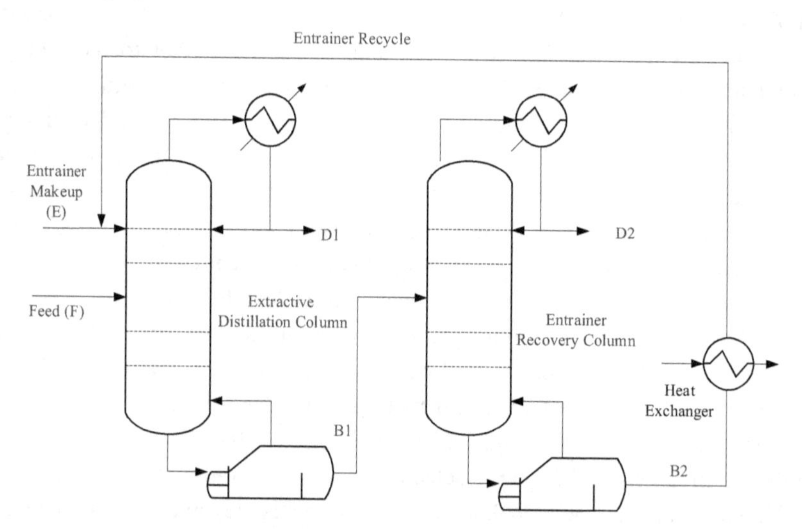

Fig. 1. Typical process flowsheet to separation methylal by the extractive distillation process.

2.1. *Residue Curve Map*

In this study, the NRTL activity model is selected as the property package in commercial software using the built-in binary interaction parameters in the simulator. Residue curve map (RCM) is used to design and analyze for distillation boundaries and tie lines in the ternary phase diagram. Tetra-ethylene-glycol (Tetra EG) is appropriate entrainer for the separation process of methylal and methanol [13]. The energy consumption in the RDC to recovery Tetra EG increase because of the high boiling point of Tetra EG.

The residue curve map of methylal/methanol/propylene-glycol system at 1 atm as showed in Fig. 2, propylene-glycol (PG), and tetra-ethylene-glycol (Tetra EG) give a good performance to modify the vapor-liquid equilibrium curve from an unstable node (the binary azeotrope) to a stable node. Although the isovolatillity line for Tetra EG (Fig. 2(b)) intersects the vertical axis quite close to methylal corner, the boiling point of Tetra EG is the higher than the boiling point of PG (Fig. 2(a)).

3. Response Surface Methodology

Response surface methodology (RSM) is one of statistical methods that can be helpful for optimization processes. RSM is applied to predict the relative significance of several variables [14].

Central composite design (CCD) is a standard RSM design that is applied to find the optimal condition in the extractive distillation column. The desired model response of the independent variables is obtained by using RSM and regression analysis. The predicted optimal condition is given

$$\hat{y} = \beta_0 + \sum_{i=1}^{k} \beta_i x_i + \sum_{i=1}^{k} \beta_{ii} x_i x_i + \sum_{i=1}^{k} \sum_{j=1, j<i}^{k} \beta_{ij} x_i x_j + \varepsilon \quad (1)$$

where $\hat{y}$ is the predicted response x_i is the levels of the independent variables, i, j, ij, and ii is the linear coefficient, the quadratic coefficient, the interaction effect, and the squared effect, respectively. Moreover, ε is the random error, and β is the regression coefficients of the independent variables [14].

The investigation of the interaction between process parameters and responses is analysed by ANOVA. The quality of the polynomial model is inspected by R^2 and P-value. Also, the operational conditions are presented in the 3 D diagram to predict the optimal conditions. To obtain the quadratic model of RSM is simulated the data following by modifying to eliminate the term of statistically insignificant. The model summary statistic demonstrates the high coefficient of determination $R^2 > 0.9$ and P-value <0.05 that suggest the model for the product quality and energy consumption of EDC.

To design the EDC, there are 6 factors (total number of stages (Nt), feed stage of mixture (NF) and entrainer (NE), temperature of entrainer, entrainer to feed ratio (E/F), and reflux ratio (Rf) that have been considered to minimize the reboiler duty and to achieve the product purity [13].

(a)

(b)

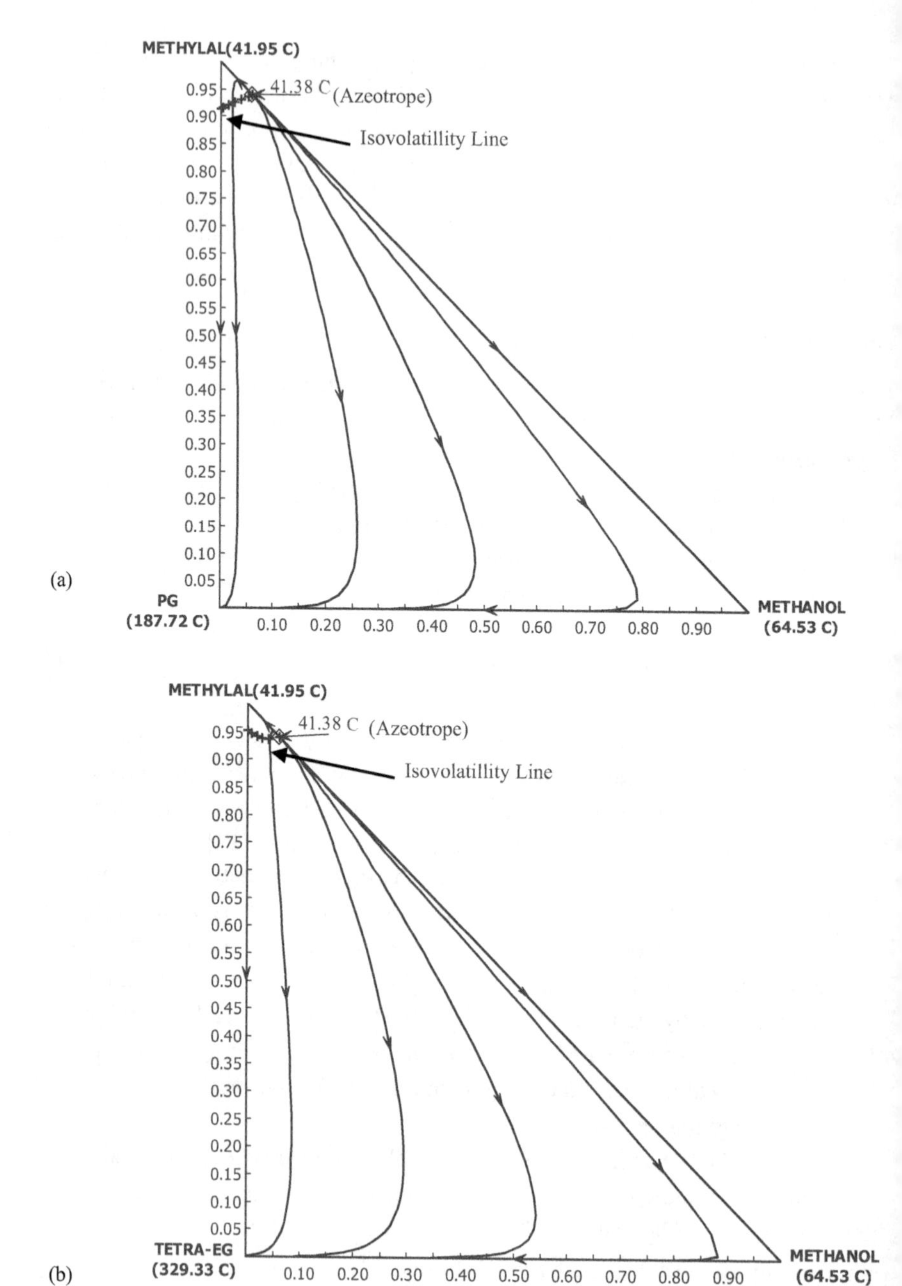

Fig. 2 Residue curve map at 1 atm for the system Methylal/Methanol/ PG (a), methylal/methanol/ TRTRA-EG (b).

4. Simulation Results and Discussion

In this paper, the excess methanol was fed into the reversible reaction to shift the equilibrium reaction to the right-hand side. Hence the mixture in the production stream combines with methanol, methylal, and water. After pretreatment, water was removed from the feed stream of the extractive distillation process. To reduce the simulation runs for 6 factors (53 runs), the design of EDC would be 2 parts; optimal structural and operational parameters. The simulation scenario can be reduce to be 20 runs in the each part that has 3 factors. In the construal of the EDC, total number of stages (Nt), feed stage of mixture (NF) and entrainer (NE) are the factor that have range of variables was given in Table 1. Moreover, Table 2 shown the range of variables for operational in EDC (temperature of entrainer, entrainer to feed ratio (E/F), and reflux ratio (Rf).

Table 1.Variables code levels for structural parameters of EDC

Variables code	Simulation range		
	-1	0	1
x_1 :Total number of stages (Nt)	9	10	11
x_2 : Feed stage of mixture (NF)	6	7	8
x_3 :Feed stage of entrainer (NE)	3	4	5

Table 2.Variables code levels for operational parameters of EDC

Variables code	Simulation range		
	-1	0	1
x_4 :Temperature of entrainer feeding (ET)	50	75	100
x_5 :Entrainer to feed ratio (E/F)	0.5	0.75	1
x_6 :Reflux ratio (Rf)	0.6	0.9	1.2

To obtain the best model, ANOVA is used to analysis. A quadratic model of RSM is applied to simulate data as followed by the eliminating the term to find statistically insignificant. The model summary statistic demonstrates the high coefficient of determination R^2 (>0.9) and a P-value lower than 0.05 to suggest the models for product purity and energy consumption of the extractive distillation column. In Table 3 Anova analysis results indicate the product purity of methylal (y_1) in EDC that was directly related to the number of stages (Nt) (x_1), location feed stage of the mixture (NF) (x_2), and location feed stage of entrainer (NE) (x_3). Moreover, the interaction between x_1 x_2 and x_1 x_3 were significant. The overall equation for the model relating the purity of methylal in EDC of code factors was given by

$$\begin{aligned} y_1 &= 0.95334 + 0.007710x_1 + 0.001528x_2 - 0.001278x_3 \ldots \\ &- 0.000614x_1^2 - 0.000614x_2^2 - 0.000489x_3^2 \ldots\ldots \\ &+ 0.000625x_1x_2 + 0.000375x_1x_3 + 0.000125x_2x_3 \end{aligned} \quad (2)$$

Table 3. Anova analysis for response of structural parameters of EDC

Source	DF	Sum Square	Mean Square	F-Value	P-Value
Model	9	0.000056	0.0000006	163.94	0.000
Linear	3	0.000034	0.000011	299.35	0.000
x_1	1	0.000028	0.000028	724.03	0.000
x_2	1	0.000002	0.000002	41.04	0.000
x_3	1	0.000005	0.000005	132.99	0.000
Square	3	0.000018	0.000006	154.17	0.000
x_1^2	1	0.000009	0.000009	248.70	0.000
x_2^2	1	0.000009	0.000009	248.70	0.000
x_3^2	1	0.000006	0.000006	157.70	0.000
Interaction	3	0.000004	0.000001	38.31	0.000
$x_1 \cdot x_2$	1	0.000003	0.000003	82.09	0.000
$x_1 \cdot x_3$	1	0.000001	0.000001	29.55	0.000
$x_2 \cdot x_3$	1	0.000000	0.000000	3.28	0.100
Error	10	0.000000	0.000000		
Lack-of-Fit	5	0.000000	0.000000	*	*
Pure Error	5	0.000000	0.000000		
Total	19	0.000057			
$R^2 = 99.33\%$		Adjust $R^2 = 98.72\%$		Pred $R^2 = 94.74\%$	

The effects of changing between x_1 and x_2 on the methylal purity as shown in the 3D map and the contour plots are presented in Fig. 3 (a) and (b), respectively. The interaction amount of x_1 and x_3 on the methylal purity in EDC were shown in Fig 4. The simulation results demonstrate that for each response, different parameters and interactions are important; for optimization of EDC, all parameters should be considered simultaneously. The optimum values of parameters are shown in Table 4 and the response for the optimal condition to predict the point option by RSM. The total of stages number, feed stage of mixture and entrainer were estimated to be 10, 7, and 4, respectively to obtain the purity of methylal at 99.90 wt. %.

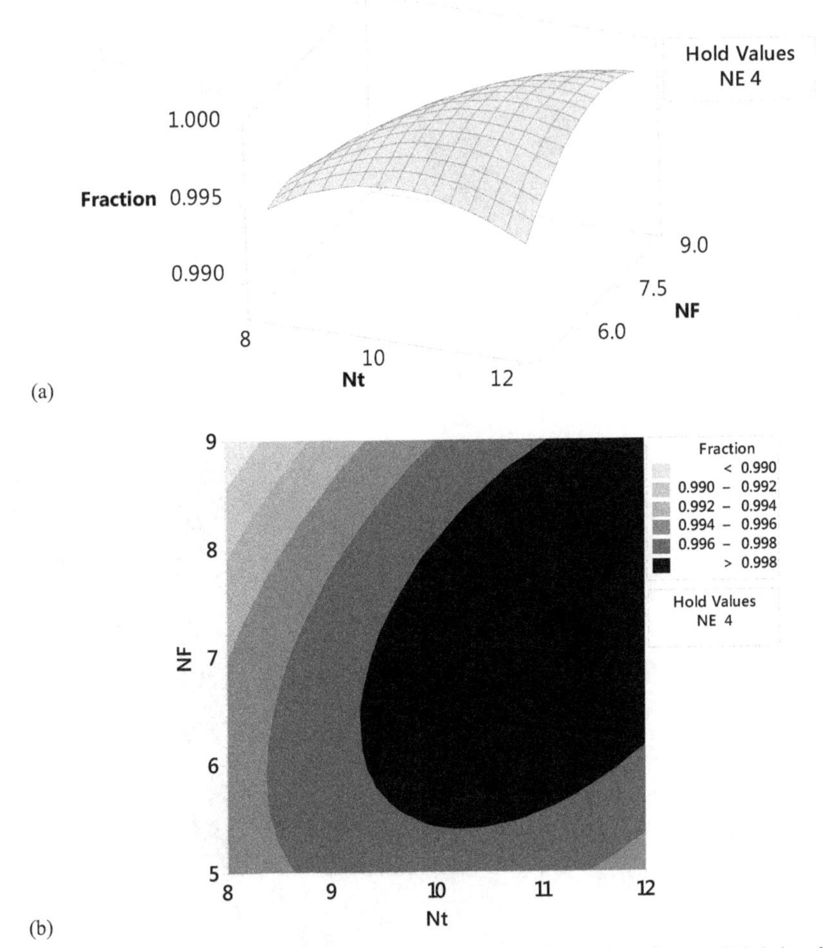

Fig. 3 Response surface for methylal purity for EDC depends on the number of stages (Nt) (x_1) and location feed stage of the mixture (NF) (x_2) for (a) 3D map and (b) contour plots.

Table 4. Optimum parameter for structural parameter of EDC

Nt: stages number	NF: feed stage of mixture	NE: feed stage of entrainer
10	7	4

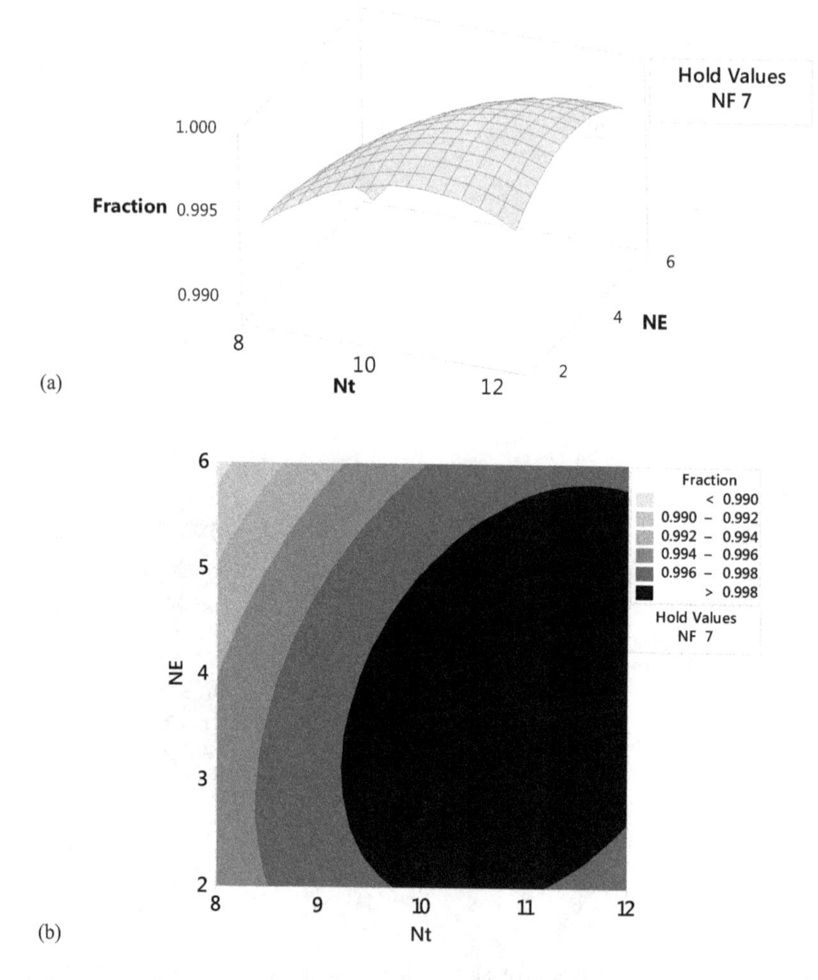

Fig. 4 Response surface for methylal purity for EDC depends on the number of stages (Nt) (x_1) and location feed stage of the entrainer (NE) (x_3) for (a) 3D map and (b) contour plots.

In Table 5 Anova analysis results indicate the energy consumption (y_2) in EDC that was directly related to the temperature of entrainer feeding (ET) (x_4), entrainer to feed ratio (E/F) (x_5), and reflux ratio (Rf) (x_6). Moreover, the interaction between x_4 x_5 was significant. The overall equation for the model relating the energy consumption in EDC of code factors was given by

$$y_2 = 35587 + 34.9x_4 + 29047x_5 + 34877x_6 - 0.2245x_4^{\ 2} \ldots$$
$$+ 0.8752x_5^{\ 2} + 31x_6^{\ 2} - 263.20x_4x_5 - 1.16x_4x_6 - 47x_5x_6 \tag{3}$$

Table 5. Anova analysis for response of operational parameters of EDC

Source	DF	Sum Square	Mean Square	F-Value	P-Value
Model	9	2,666,386,549	296,265,172	17,343.58	0.000
Linear	3	2,635,336,248	878,445,416	51,424.83	0.000
x_4	1	389,05,6886	38,905,6886	22,775.67	0.000
x_5	1	501,427,415	501,427,415	29,353.92	0.000
x_6	1	1,744,851,947	1,744,851,947	102,144.89	0.000
Square	3	9,401,061	3,133,687	183.45	0.000
x_4^2	1	495,115	495,115	28.98	0.000
x_5^2	1	7,523,563	7,523,563	440.43	0.000
x_6^2	1	193	193	0.01	0.917
Interaction	3	21,649,240	7,216,413	422.45	0.000
$x_4 \cdot x_5$	1	21,648,529	2,1648,529	1,267.32	0.000
$x_4 \cdot x_6$	1	611	611	0.04	0.854
$x_5 \cdot x_6$	1	100	100	0.01	0.940
Error	10	170,821	17,082		
Lack-of-Fit	5	17,0821	34,164	*	*
Pure Error	5	0	0		
Total	19	2,666,557,370			
$R^2 = 99.99\%$		Adjust $R^2 = 99.99\%$		Pred $R^2 = 99.95\%$	

The effects of changing between temperature of entrainer feeding (x_4) and entrainer to feed ratio (E/F) (x_5)on the energy consumption in EDC as shown in the 3D map and the contour plots are presented in Fig 5 (a) and (b), respectively. Results demonstrate that for each response, different parameters and interactions are important. The optimal condition of operational factors can be found the optimization of energy consumption and product purity by using the point prediction of RSM. The optimum values of parameters are shown in Table 6. The optimized temperature of entrainer feeding was 125 °C, which the entrainer to feed ratio at 0.5732. The reflux ratio was operated 0.9545. Moreover, the simulation results showed that the purity of methylal could be achieve at 99.9 wt% with heat duty in the reboiler at 15.91 kW in the EDC and more detail in the process as presented in Table 7.

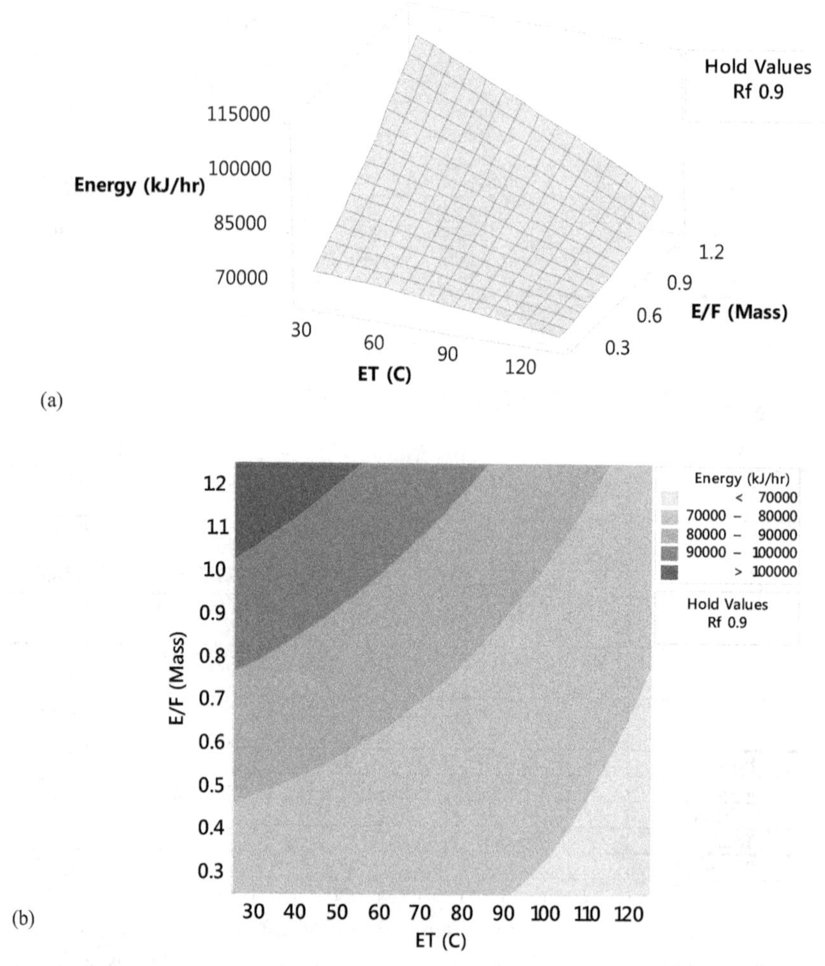

Fig. 5 Response surface for energy consumption of EDC depends on temperature of entrainer feeding (x_4) and entrainer to feed ratio (E/F) (x_5) for (a) 3D map and (b) contour plots.

Table 6. Optimum parameter for operational parameter of EDC

ET: Entrainer Temperature	E/F: Entrainer to feed ratio	Rf: Reflux ratio
125 °C	0.5732	0.9545

Table 7. Optimized specifications and operating conditions for methylal/methanol separation using PG as entrainer

Column	**Extractive distillation column (EDC)**	
	Pressure (atm)	1
	Total stages	10
	Feed stages	7
	Entrainer feed stages	4
	Mass reflux ratio	0.9545
	Distillate rate (D1) (kg/h)	94
	Entrainer recovery column (RDC)	
	Pressure (atm)	1
	Total stages	5
	Feed stages	3
	Mass reflux ratio	1
	Bottom rate (B2) (kg/h)	57.3
Streams	**Feed stream (F)**	
	Temperature (°C)	25
	Pressure (atm)	1
	Feed flow rate (kg/h)	100
	Entrainer stream (E)	
	Temperature (°C)	125
	Pressure (atm)	1
	Feed flow rate (kg/h)	57.32
	D1 stream	
	Temperature (°C)	41.9943
	Mass fraction	
	- Methylal	0.999
	- Methanol	0.001
	D2 stream	
	Temperature (°C)	107.944
	Mass fraction	
	- Methylal	0.59
	- Methanol	0.401
	- PG	0.009
Heat duty	**Extractive distillation column (EDC)**	
	- Condenser (kW)	-18.98
	- Reboiler (kW)	19.51
	Entrainer recovery column (RDC)	
	- Condenser (kW)	-2.39
	- Reboiler (kW)	5.86
	Heat exchanger (kW)	-2.98

5. Conclusions

In this work, the separation of methylal and methanol by extractive distillation was intensified by PG as entrainer. For this system, NRTL model in the commercial simulator is used to simulate. In the residue curve map (RCM) of the methylal / methanol / PG as shown that PG is possible to obtain high purity of methylal in the extractive distillation process. A practical method to propose the design of the extractive distillation column (EDC) has been obtained by response surface methodology (RSM) with the study of the effect of factors and their interactions on the requirement of energy consumption and methylal purity. The optimized parameters for EDC design by using RSM can reduce simulation scenarios by separate for 2 parts. Total number of stages, feed stage of mixture and entrainer (NE), entrainer temperature, entrainer to feed ratio, and reflux ratio (Rf) are directly related to product purity and energy consumption. To obtain the optimal operating conditions, sensitivity analysis of the EDC has been studied by RSM that is sufficient and efficient for design the extractive distillation system.

Acknowledgments

The authors appreciatively acknowledge the financial support for the Research Council of Naresuan University (R2562E023), and also the authors are thankful of Chulalongkorn University (Department of Chemical Engineering) for providing a place to research.

References

1. E. Carretier, Ph. Moulin, M. Beaujean, and F. Charbit, "Purification and dehydration of methylal by pervaporation," *Journal of Membrane Science*, **217**, pp.159-171 (2013).
2. J.P.Knapp, and M.F. Doherty, "A New Pressure swing distillation process for separating Homogeneous azeotropic mixture," *Ind. Eng. Chem. Res.*, **31**, pp. 346-357 (1992).
3. B. Yu, Q.Wangn and C. Xu, Design and control of distillation system for methylal/methanol separation. Part 2: pressure swing distillation with full heat integration. *Ind. Eng. Chem. Res.,* **51**, pp.1293–1310 (2012).
4. P. Kittisupakorn, K. Jariyaboon, and W. Weerachaipichasgul, "Optimal high purity acetone production in a batch extractive distillation column," *Lecture Notes in Engineering and Computer Science: Proceedings of the International MultiConference of Engineers and Computer Scientists 2013, IMECS 2013, March, 2013,* Hong Kong, pp. 143-147.
5. Y. Ren, Z. Huang, H. Miao, Y. Di, D. Jiang, K. Zeng , L. Bing, and W. Xibin, "Combustion and emissions of a DI diesel engine fuelled with diesel-oxygenate blends," *Fuel*, **88**, pp. 2691-2697 (2008).
6. E. Hu, Z. Gao, Y. Liu, G. Yin, and Z. Huang, "Experimental and modeling study on ignition delay times of dimethoxy methane/n-heptane blends," *Fuel*, **189**, 350-357 (2016).
7. S. Satoh, and T. Yukio, "Process for Producing Methylal," *U.S. Patent 6,379,507*, April 30 (2002).

8. G.P. Hagen, and M.J. Spangler, "Preparation of polyoxymethylene di-alkane ethers, by catalytic conversion of formaldehyde formed by dehydrogenation of methanol or dimethyl ether," *U.S. Patent 6,350,919*, February 26 (2002).
9. X. Zhang, S. Zhang, and C. Jian, "Synthesis of methylal by catalytic distillation," *Chemical Engineering Research and Design*, **89**, pp. 573-580 (2011).
10. Y. Dong, C. Dai, and Z. Lei, "Extractive distillation of methylal/methanol mixture using the mixture of dimethylformamide (DMF) and ionic liquid as entrainers," *Fuel*, **216**, pp.503-512 (2018).
11. Y. Dong, C. Dai and Z. Lei, "Extractive distillation of methylal/methanol mixture using the ethylene glycol as entrainer," *Fluid Phase Equilibria*, **462**, pp. 172-180 (2018).
12. M.A.S.S. Ravagnani, M.H.M. Reis, F.R. Maciel, and M.R. Wolf-Maciel, "Anhydrous ethanol production by extractive distillation: A solvent case study," *Process Safety and Environmental Protection,* **88**, pp.67-73 (2010).
13. W. Weerachaipichasgul, A. Chanpirak, and P. Kittisupakorn, "Response surface methodology in optimization of separation process for methylal /methanol based on process simulation of extractive distillation," *Lecture Notes in Engineering and Computer Science: Proceedings of The International MultiConference of Engineers and Computer Scientists 2019, IMECS 2019, 13-15 March, 2019,* Hong Kong, pp. 431-435.
14. S. L. Mitra, R. Rahbar, and Z. Mortaza, "Response surface methodology in optimization of a divided wall column," *Korean J. Chem. Eng.*, pp.1414-1422 (2018).

Update to Strong and Weak Force Proposed Equations*

Matthew Cepkauskas,
Baker Hughes, a GE Company, Jacksonville, Florida, USA
E-mail: mattcep1@gmail.com

The search for grand unification, the understanding and identifying the relations of forces of physics, has been the interest of many particle physicist for many decades. The present paper investigates the validity of a previous paper that proposed two possible equations for both the strong force and the weak force. It investigates the magnitude of these equations for the Hydrogen atom. Also, the coupling constant are determined and compared with known values. Both the forces and coupling constants show good agreement with existing known results. Other elements on the periodic table are used to further evaluate the validity of the equations. It is also shown that the four forces of strong, weak, gravitational and Coulomb converge to the expected energy and corresponding length identified by the Standard model and Super Symmetry models. It is believed that these equations represent the dominate terms in a perturbation series solution. The work is extended to include a strong force pressure with comparison to recent data and single degree of freedom frequency equations for each of the four forces. In addition periodic table numerical results for pressure and frequencies are included.

Keywords: Strong; Weak; Electro Magnetic and Gravitational Forces; Grand Unification.

1. Introduction

The search for grand unification, the understanding and identifying the relations of forces of physics, has been the interest of many particle physicist for many decades [1]. The present investigation is a continuation of a previous paper found in [2] to determine the validity of proposed potential weak force and strong force equations.

In grand unification, two of the four forces are well known namely Newton's gravity and Coulomb's electric charge, the second two forces, weak and strong, do not have known force equations associated with them, thus the four equations can be written as:

$$F_{Gravity} = G\frac{m^2}{L^2} \tag{1-a}$$

*This work is self-supported.

$$F_{Coulomb} = K_e \frac{Q_0^2}{L^2} \tag{1-b}$$

$$F_{Weak} = ?? \tag{1-c}$$

$$F_{Strong} = ?? \tag{1-d}$$

In [2], using the pi theorem for dimensional analysis, the following five forces (gravity, Coulomb, thermodynamic, quantum and relativity) were identified. These equations are well known, and the dimensions can simply be confirmed as correct without using the pi theorem.

$$F_{Gravity} = G \frac{m^2}{L^2} \tag{2-a}$$

$$F_{Coulomb} = K_e \frac{Q_0^2}{L^2} \tag{2-b}$$

$$F_{Thermo} = K_b \frac{T}{L} \tag{2-c}$$

$$F_{Quantuum} = \bar{h} \frac{1}{tL} \tag{2-d}$$

$$F_{Relativity} = c \frac{m}{t} \tag{2-e}$$

Comparison of these two sets of equations begs the question: Is the weak and strong force a combination of the five forces? This question was the basis for [2].

2. Proposed Equations

2.1. *Background*

As reported in [2], today it is understood that the approximate ratio of forces when normalized to the force of gravity is as follows:

Gravity 1.0
Weak 1.0E25
Electromagnetic 1.0E36
Strong 1.0E38

In [2], 45 cases of permutation of possible products of the five equations were investigated using the parameters of interest [mass, length, temperature, time and charge, along with gravitational, Coulomb, Boltzmann, Planck and relativity constants] for the Hydrogen atom. This resulted in the identity of two potential weak equations and two potential strong force equations that when normalized to gravity agreed with the known weak and strong values, above. An error was made in [2], the size of the hydrogen atom was incorrectly taken as 1.0 E-15 meters. When the size is corrected to 1.0 E-10 meters, the following equations were found:

Potential Weak Force Equation with Coupling Constants:

$$F_{weak\_1} = \frac{F_{Thermodynamic} F_{Quantum}}{F_{Relativity}} = \frac{\bar{h} K_b}{c} \frac{T}{L^2 m} \quad \text{(3-a)}$$

$$\alpha_{weak\_1} = \frac{K_b}{c^2} \frac{T}{m} \quad \text{(3-b)}$$

$$F_{weak\_2} = \frac{F_{Coulomb} F_{Quantum}}{F_{Relativity}} = \frac{\bar{h} K_e}{c} \frac{Q_0^2}{L^3 m} \quad \text{(4-a)}$$

$$\alpha_{weak\_2} = \frac{K_e}{c^2 L} \frac{Q_0^2}{m} \quad \text{(4-b)}$$

Potential Strong Force Equation with Coupling Constants:

$$F_{strong\_1} = \frac{F_{Relativity} F_{Coulomb}}{F_{Quantum}} = \frac{cmK_e}{\bar{h}} \frac{Q_0^2}{L} \quad \text{(5-a)}$$

$$\alpha_{strong\_1} = \frac{mLK_e}{\bar{h}^2} Q_0^2 \quad \text{(5-b)}$$

$$F_{strong\_2} = \frac{F_{Thermodynamic} F_{Relativitty}}{F_{Quantum}} = \frac{cmK_b T}{\bar{h}} \quad (6\text{-a})$$

$$\alpha_{strong\_2} = \frac{K_b TmL^2}{\bar{h}^2} \quad (6\text{-b})$$

The incorrect diameter used in reference [2] was fortuitous in that it showed that these equations can either be weak or strong force equations. Note that not only are 3 of the 4 equations dependent on the length dimension, but also since they are normalized to the gravitational force (also dependent on the length) all four equations are influenced by the choice of length. It is well known that both the weak and strong force are dependent on distance, thus strong_2 equation, 6-a & 6-b is eliminated from consideration. This identified the need to examine these forces further from the size of the hydrogen atom to much smaller dimensions up to 10 E-23 meters. This effort identified the best choice of the possible equations to be:

2.2. *Potential Weak Force Equation with Coupling Constants*

$$F_{Weak} = \frac{F_{Relativity} F_{Coulomb}}{F_{Quantum}} = \frac{cmK_e}{\bar{h}} \frac{Q_0^2}{L} \quad (7\text{-a})$$

$$\alpha_{Weak} = \frac{mLK_e Q_0^2}{\bar{h}^2} \quad (7\text{-b})$$

2.3. *Potential Strong Force Equation with Coupling Constants*

$$F_{Strong} = \frac{F_{Coulomb} F_{Quantum}}{F_{Relativity}} = \frac{\bar{h}K_e}{c} \frac{Q_0^2}{L^3 m} \quad (8\text{-a})$$

$$\alpha_{Strong} = \frac{K_e Q_0^2}{c^2 L m} \quad (8\text{-b})$$

3. Results

Rohlf [5] provides a plot of relative strength of the four forces in his Figure 1-19. Plotting equations 2-a, 2-b, 7-a and 8-a and taking the same liberty as found in [5] where the strong force is negligible for a dimension greater than 10 E-15 m and the weak force is negligible for a dimension greater than 10-18 m results in figure 1. Figure 1 shows the gravity force for electron-electron, electron-proton and proton- proton for clarity. The resemblance to the results found in [5] is remarkable.

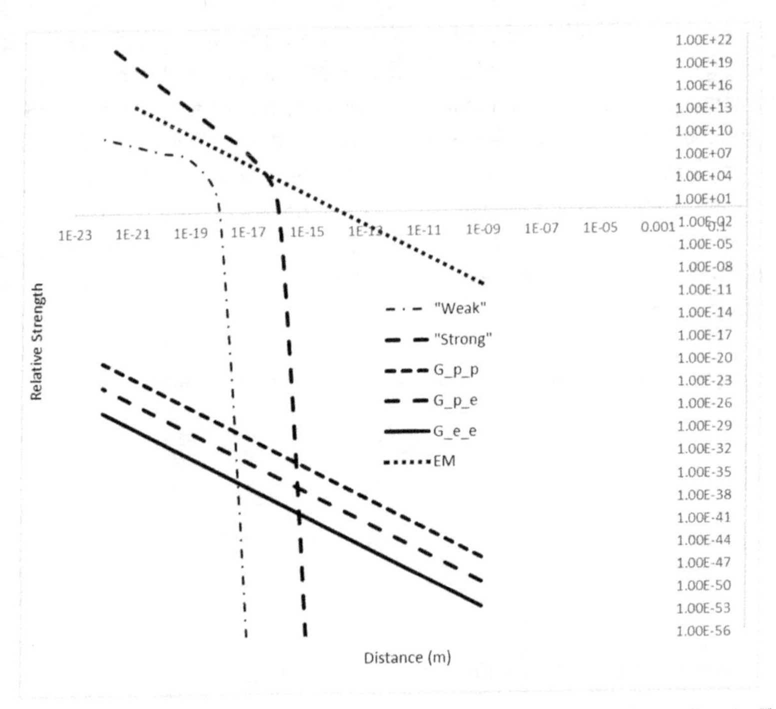

Fig. 1 Relative Strength of Weak, Electromagnetic, Strong and Gravitational forces. Gravity Shown for Proton-Proton, Proton-Electron and Electron-Electron Combinations

Per Reference [5], the coupling constants are given with numerical values of:

Strong coupling constant = 1.0
Electro Magnetic coupling constant = 1/137
Weak Coupling constant = 10-6 to 10-7
Gravity Coupling constant = 10-39

The proposed equations have the coupling constants for strong and weak force as being a function of dimensions resulting in the weak force ranging from 3.5 E-5 at 10 E-18 m to 3.5 E-9 at 10 E-22 m and the strong force starting at 0.0015 at 10 E-15 m to 1.5 E+4 at 10 E-22 m for Hydrogen. This is in agreement with the expected coupling constants found in [5].

The coupling constant equations for the Coulomb, Weak and Strong force are plotted versus electron volts by converting length to electron volts, by use of dimensional analysis. It should also be noted that mass and charge can be converted to electron volts. Figure 2 and 3 give results using Planck's dimensions as defined in [2] with the proposed strong and weak force. This represents a check on the dimensional conversion as the forces in Figure 2 converge to Planck's force of 1.21xE+44 Newtons at Planck's length of 1.616x10-35meters. Figure 3 shows the convergence to exist at Planck's energy of 1.22x1019 GeV.

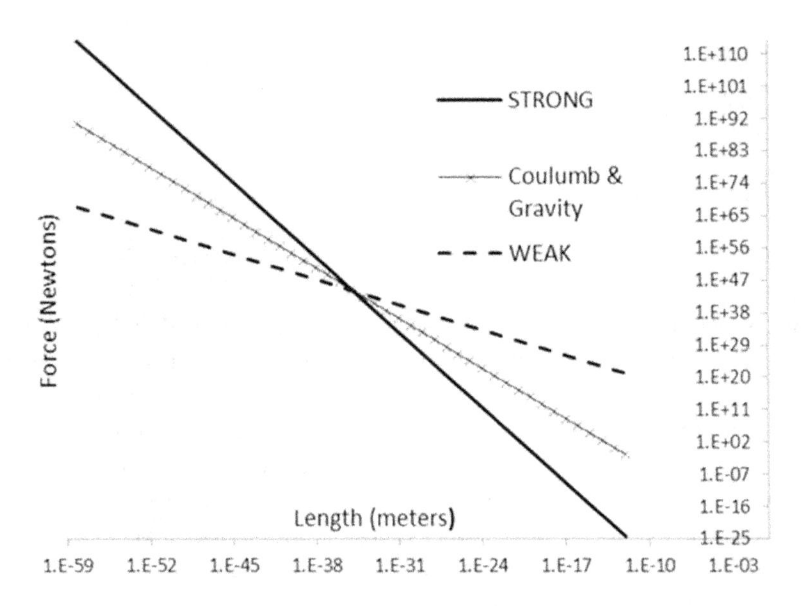

Fig. 2 Planck's Force Versus Length

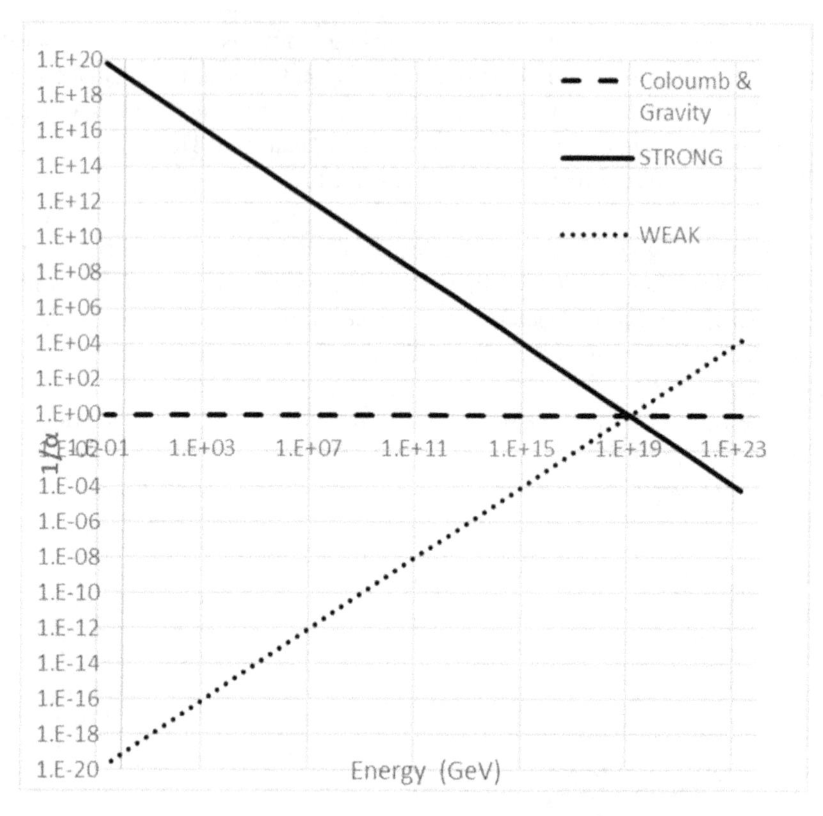

Fig. 3 Planck's Inverse Coupling Constants versus Energy

By maintaining the hydrogen charge and increasing the hydrogen mass to 15 orders of magnitude to a mass of 1.67 E-12 Kg, results in behavior at high energy levels, known as the Grand Unified Theory (GUT). Figure 4 shows, using the proposed equations with gravity and Coulomb for the inverse of the coupling constants, convergence of three of the four forces occurs at 10E+16 GeV (10E-32 meters). By increasing the mass to 1.67 E-9 Kg results in convergence of all four forces at an energy level of ~10E18 GeV, the energy level of the Theory of Everything (TOE), as found in physics texts [3,4,5] based on the standard model and supper symmetry.

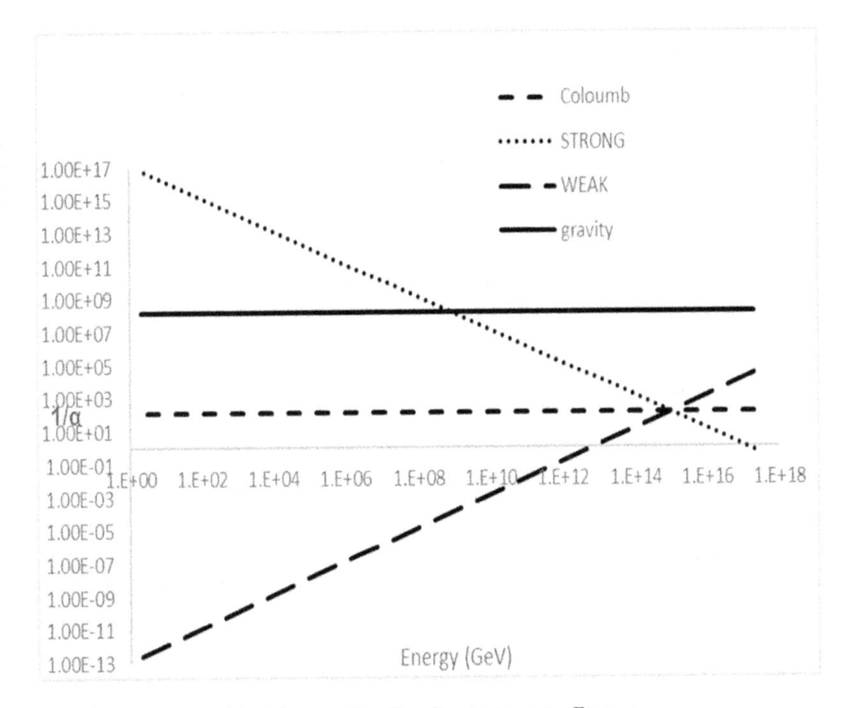

Fig. 4 Inverse Coupling Constants versus Energy

In addition, the weak and strong force are shown in Figure 5 and Figure 6 for other elements on the periodic table with break in continuity going from one row of the periodic table to the next. Figure 6 shows an increase in strong force with increasing number of protons as expected. Figure 5 also shows an increase in weak force. It is not clear that Figure 5 supports the proposed weak equation, a force that would bring an unstable element to equilibrium but is presented here for completeness.

Figure 5 & 6 need a word of caution. It appears that the strong force is smaller than the weak force. The reason for this is the diameter was taken as twice the atomic radius, which is on the order of 10 E-10 m, but the weak and strong force only exhibit significance in distances less than 10 E-15. As the dimensions get smaller the order of magnitude of the strong and weak equations reverse. Thus figure 5 & 6 should be individually looked at as relative behavior within the periodic table and not compared to each other.

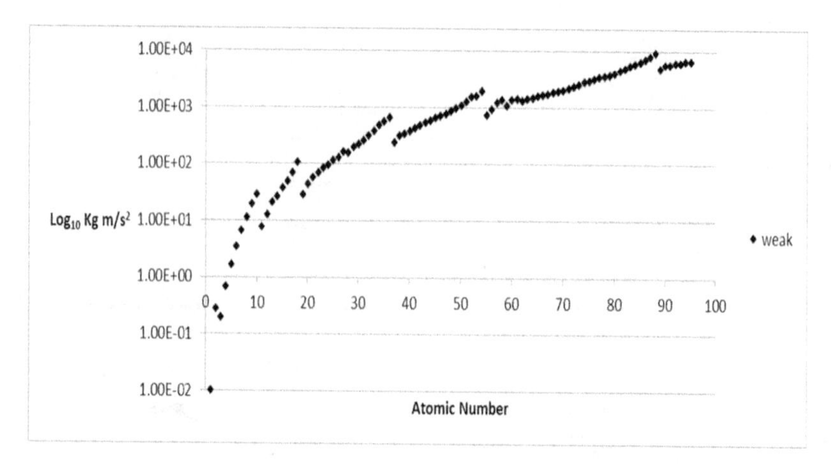

Fig. 5 Weak Force for Elements on the Periodic Table

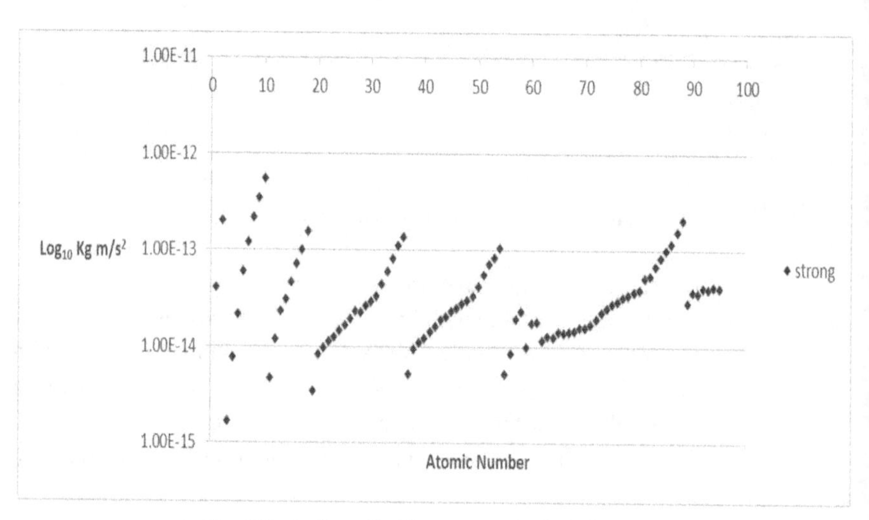

Fig. 6 Strong Force for Elements on the Periodic Table

□

4. Discussion

The simple equations proposed for the strong force and weak force leaves the author in a quandary. Can it be so simple? Are these equations a valid approximation or at least a possible teaching tool? Does one not report the results for fear of errors in the logic? A critical review of the approach by the Physics community is appreciated.

Note that equating the coupling constants result in an equation at which GUT converges. That is $mL=\hbar/c$. This same identity results are obtained by equating the Coulomb coupling constant with the Strong coupling constant and then to the weak coupling constant. A similar expression for TOE would be to equate the gravity coupling constant with the Coulomb resulting in $m_1m_2/Q^2 = Ke/G$. This convergence is not surprising as indicated in [4], two non-parallel lines will converge.

The logic utilized in [2] to obtain the proposed equations consisted of:
1. Treat the available relative strength information as experimental test results.
2. As in many test, use dimensional analysis to reduce the number of parameters.
3. Speculate if the strong and weak force can be a product of the five fundamental forces (denote them as subscripts 1,2,3,4,5 in lieu of thermal, quantum, relativity, Coulomb and gravity).
4. Look for products such as F1*F2/F3 or F1*F2*F3/F4*F5 (45 cases in total) that when normalized to gravity match the available relative strength data.
5. One possible flaw maybe in the choice of exponents for the forces. It was noted that in Planck's units the force exponent would all need to be one since all forces are equal as shown in [2]. However, the data that was being used was based on today's relative strength. Thus, it was assumed that today similar distribution, i.e. exponent of one is correct.

Although the two equations, 3 and 6, that are a function of Thermodynamic, Quantum and Relativity forces were not investigated in detail, two observations are made:

1. Both the identified strong force equations (5-a and 6-a) and weak force equations (3-a and 4-a) when equated yield: $TL=q^2K_e/K_b$ which identifies a relationship between temperature and charge, such an identity does exist in Planck units. Proof: Consider $L_P*T_P=SQRT(\hbar c^5/G)*SQRT(G\hbar/c^3)/K_b=c\hbar/K_b=q_P{}^2K_e/K_b$, therefore $L_P*T_P = q_P{}^2K_e/K_b$.
2. Since for example, equation 5-a and equation 6-a are on the same order of magnitude, their sum is of a similar order of magnitude and also qualifies for a possible combination to make up the strong force

5. Update

To this point the paper has not been altered and appears exactly as found in [7] with the exception that Figures 2, 3 and 4 have been corrected. They were originally labelled backwards, STRONG FORCE should have read WEAK

FORCE and vice versa. They are now correct in this manuscript. Recently, [8] reported a measured predicted average pressure of 10^{35} Pascals inside a proton. The proposed strong force equation divided by dimension of length square predicts a pressure on the order of 10^{36} Pascals at L= .1 E-15 and 10^{30} Pascals at L=2 E-15 for Hydrogen. Pressures for the periodic table, assuming the proton is proportional to the size of the atom, but five orders of magnitude lower, i.e. 1.0E-15 are given in Figure 7.

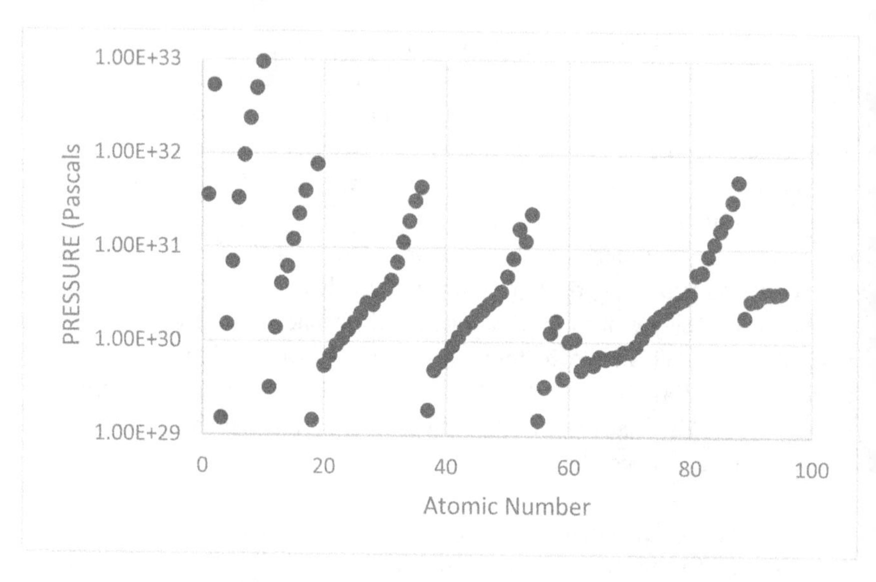

Fig. 7 Strong Force Pressure normalized to E-15

The authors original interest, as explained in [8], in this subject was related to a simple spring mass, the equation appears in much of applied mechanics and also in physics, so the original search was to find four simple spring mass equations that would be representative of the four forces. Knowing the Force and dividing by the length results in a spring constant for each force and knowing the mass the frequency is obtained. It should be noted that if unification occurs with the forces being equal, then also the frequencies should be equal, thus unification would appear to be a resonance type of condition. The four forces would then have a related frequency of:

$$f_{n\_weak} = \frac{Q_0}{2\pi L}\sqrt{\frac{cK_e}{\hbar}} \tag{9-a}$$

$$f_{n\_strong} = \frac{Q_0}{2\pi m L^2}\sqrt{\frac{\hbar K_e}{c}} \tag{9-b}$$

$$f_{n\_gravity} = \frac{1}{2\pi L}\sqrt{\frac{Gm}{L}} \tag{9-c}$$

$$f_{n\_EM} = \frac{Q_0}{2\pi L}\sqrt{\frac{K_e}{Lm}} \tag{9-d}$$

Using 1.0 E-18 as a distance that the thee forces EM, weak and strong would interact, the frequencies for these three forces are given in Figure 8. The fourth row of the periodic table, consisting of stable metals, appears to have natural frequencies close together.

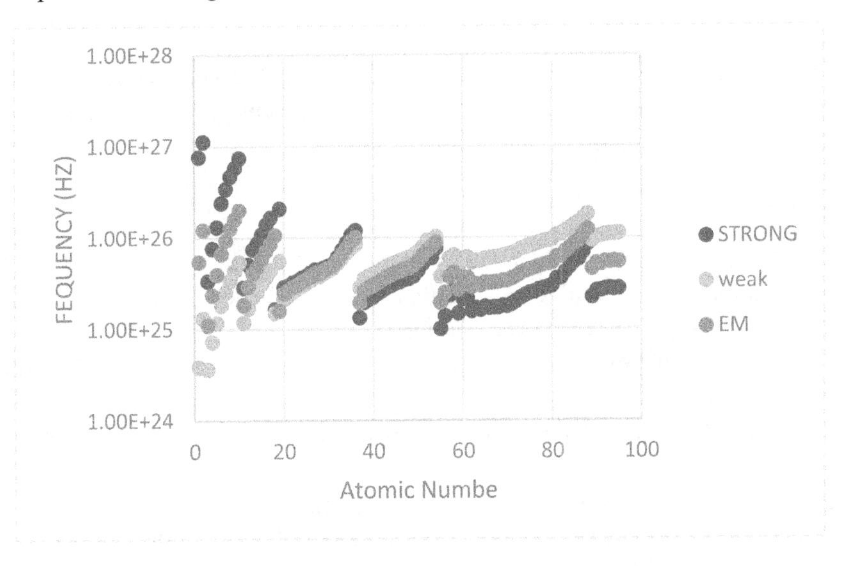

Fig. 8 Natural Frequencies for strong, weak and EM at L= 1.0E-18

6. Conclusion

The present paper has identified a possible force equation for the weak and strong force with supporting data for the hydrogen atom and convergence of forces. One would not expect these equations to be exact but may represent the dominant term in a series solution. It also shows the behavior of these equations for elements of the periodic table.

An additional comparison has been made with literature regarding the pressure in a proton, along with the pressure for the periodic table of elements assuming the proton size is proportional to the size of the atom. Natural frequency equations based on dimensional analysis associated with the different forces were provided along with a periodic table comparison for the example of the frequencies at 1.0E-18m.

On one hand the results appear too simple to be correct, yet on the other hand these simple equations appear to have some of the required ingredients and behave in a manner suggested by more sophisticated models. It should be noted the equations are limited to the original dimensional analysis assumption as being dependent on five physical constants and five parameters of time, temperature, charge, mass and length. Thus the results is void of any statistical probability or higher level mathematics as found in QED or QCD.

7. Acknowledgments

My renaissance wife of forty-eight years, without her continuous help and encouragement this endeavor would not occur. This subject was first introduced to me some 45 years ago by author AY of [1].

References

1. Paul H. Frampton, Sheldon L. Glashow, Asim Yildiz.
2. Math Sci Press, 1980 - Grand Unified Theories (Nuclear physics).
3. Cepkauskas, M.M., "Planck's Units Re-visited", 12th Pan-American Congress of Applied Mechanics, January 02-06, 2012, Port of Spain, Trinidad.
4. David J. Griffiths, Introduction to Elementary Particles, John Wiley and Sons, Incorporated.
5. Mark Thomson, Modern Particle Physics, Cambridge University Press.
6. James William Rohlf. Modern Physics from αtoZ0, John Wiley and Sons.
7. Matthew Cepkauskas, "Strong and Weak Force Proposed Equations", "Lecture Notes in Engineering and Computer Science: Proceedings of The International MultiConference of Engineers and Computer Scientists 2019, IMECS 2019, 13-15 March, 2019, Hong Kong, pp 326-329.
8. Berkert et. al., https://www.nature.com/articles/s41586-018-0060-z
9. An Engineering Approach to Particle Physics, New Developments in Structural Engineering and Construction, Yazdani, S. and Singh, A. (Eds.), ISEC-7, Honolulu, June 18–23, 2013.

Effect of Bandgap Vs Electron Tunneling on Photovoltaic Performance of Ringwood (Syzgium Anisatum) Dye-sensitized Solar Cell*

T.J. Abodunrin,† O.O. Ajayi, M.E. Emetere, A.P.I. Popoola, U.O. Uyor and O. Popoola
Department of Physics, Covenant University, Ota, Ogun State, P.M.B 1023, Nigeria
†E-mail: Temitope.abodunrin@univeristy.edu.ng
www.university_temitope.abodunrin.edu

O. Popoola
Tshwane University of Technology, Department of Electrical Engineering Pretoria, South Africa
E-mail: walepopos@gmail.com

The quest for more energy remains vital to energy sustainability. In the wake of several adverse consequences of indiscriminate combustion of fossil fuel, there is an urgency to exploit our natural environment for ecologically benign alternatives. This search led to *S. anisatum* dye being investigated for its prospective application in dye-sensitized solar cells. Phytochemical screening revealed the presence of phenols, flavonoids, tannins, glycosides, terpenoids and protein, presenting a wide repertoire of chromophore selection for charge transport. UV/VIS spectroscopy showed that *S.anisatum* dye exhibits multiple peak absorbances of radiation within the near ultraviolet and the visible region of the electromagnetic spectrum of light. The consequent adsorption of its chromophores into titanium structure, amplitude of inter-atomic vibrations of the novel structures formed in *S. anisatum* dye were subject of our investigation. The outcome shows TiO_2/*S.anisatum* dye interface reveals the impact of orientation on output photovoltaic performance of *S.anisatum* dye-sensitized solar cells with a short circuit current of 0.07 mA, open circuit voltage of 68.8 mV, fill factor value of 0.84 and the output efficiency was 0.027 % using KBr electrolyte. This is a comparatively good result considering previous records of dye-sensitized solar cell photovoltaic performance. The significance of these results was re-analyzed from molecular perspective with the aid of scanning electron microscopy (SEM). A need to boost the efficiency necessitated the interpretation of SEM micrographs using Gwwydion software program. These presented possible areas for charge transport within the electron shells of *S.anisatum* dye nanocomposite, and regions where tunneling occurred providing a much needed insight for future studies. Consequently, this study was expanded to accommodate the influence of bandgap on electron occupancy in *S.anisatum* shells. This elucidation captures the molecular dynamics of charge transport versus tunneling, consequent upon output performance of dye-sensitized solar cell technology explained with quantum principles. The application of

* This work is supported by Covenant University and Tshwane University of Technology.

this work is particularly relevant in modelling, photovoltaic simulations and building energy efficient models.

Keywords: Bandgap; Energy Efficiency; Energy Harvesting; Electron Tunneling.

1. Introduction

Energy statistics have revealed that power released by the sun in just an hour is sufficient to meet all mankind's energy needs annually [1]. A pertinent enquiry remains deploying the appropriate technological device to adequately harvest this energy crop for solving the growing needs of the teeming global population. A lot of research effort has been geared towards production and refinement of several technologies. The history of solar technology began from mimicry of biosynthetic process in plants, this led to the consequent manufacture of silicon solar panels [2]. Till date, silicon solar cells dominate the photovoltaic market. Their efficiency of 25% remains a very attractive feature despite a comparatively high cost of installation, inability to withstand extreme temperature conditions, undeveloped methods of disposal of expired solar panels and a complicated manufacturing process [3]. These limitations prompted several studies aimed at encrypting silicon solar cells yet obtaining salient photovoltaic efficiency output. This brought to the fore, an era of thin film production. They satisfied the basic requirement, they were cheaper photovoltaics but, lacked the high-efficiency-characteristic of silicon solar cells. This is because, the films were cheap however, the equipment for depositing the films were quite expensive. In practice, the greater the accuracy, the higher the cost of equipment. In addition, technical know-how of operating this equipment was required for optimal performance [4]. It therefore seemed a breakthrough was in sight with the advent of dye-sensitized solar cell technology. They were distinguished by a low-cost price, uncomplicated process of manufacture, readily available raw materials, ability to withstand high temperatures and operation under conditions of diverse lighting, required no special disposal system and ecologically benign [5]. This made dye-sensitized solar cells a prospective viable substitute for silicon solar cells but for the persistent challenge of poor efficiency and long- term stability [6]. A solution to the problem of poor efficiency came through the discovery of perovskite solar cells. Perovskite solar cells have recorded an efficiency of 18% with a promise of better output performance if the challenge of rapid deterioration in air and unfavorable redox reaction are addressed [7]. This buttressed the correlation between microscopic orientation, interfacial adhesion structures and macroscopic photovoltaic performances [8]. Several research efforts explored suppression of electron-hole recombination in DSCs by altering the electrolyte size, addition of additives or changing the pH of the electrolytic solution, results obtained showed a direct modification in electron dynamics that improved output efficiency by a factor of 0.4 [9]. This is because reduction in grain size created a larger surface area for adsorption and different anchoring modes with the TiO_2 frame. In effect, transformation of unstable dye aggregation to stable conformation was found to occur in the N719 dyes that were subject of consideration. Their smaller composite sizes enhanced multiple adhesion onto the TiO_2 surface, which limited electron-hole recombination by producing a layer of insulated barrier [10]. This good result encouraged more research endeavors such as insertion of phenolic rings to modify dye molecular structures, the outcome was reduction in recombination by approximately five times. A probe into this phenomenon revealed that, the five times fluctuation in recombination rate was as a result of an extension of the distance of the back-electron transfer route, which was obtained from the insertion of the dye, by means of employing quantum chemical simulations [11]. Other researchers used simplified models of some dyes to investigate the extent of structural modification on charge recombination. Their result showed that, insertion of an additional phenyl ring close to the anchoring group brought about four times reduction in the electron-hole recombination rate. Consequently, charge-transfer distance was found to be a major contributor to recombination time via theoretical analysis [12]. In view of this, present day energy research focus is directed at investigating suitable properties in dye-sensitized solar cells which may be incorporated into perovskite structures to further enhance its present efficiency. This would usher in future trends in the photovoltaic field. The incentive for this work lies in exploring some of the photovoltaic properties already discussed, applying them to *S.anisatum* DSC with a view of studying their impact on the performance, with an ultimate objective of re-branding them in the field of optoelectronics. Future research would benefit from examining co-sensitization of *S.anisatum* dye with other dyes incorporated on perovskite frames as a means of boosting the present output efficiency of dye-sensitized solar cells.

2. Materials and Method

The procedure of preparation of *S.anisatum* dye is as described in subsequent subsections.

2.1. *Dye Extraction*

200 grams of *S.anisatum* leaves were harvested and air dried until they assumed constant weight. The process of dye extraction followed standard procedures in methods [13]. The leaves were milled and spread out for some hours to eliminate vapor and discourage fungal growth. The dried coarse leaf flakes were then soaked using 2.5 litres of methanol in thin layer chromatography tanks to extract the dye. The pH of the solution was determined, the methanolic mixture was separated using a Stuart RE 300 rotary evaporator to obtain Ringwood dye.

2.2. *Phytochemical Analysis*

The phytochemical screening of *S.anisatum* was carried out under similar laboratory processes described by Najm et al., 2017. 1 g of *S.anisatum* was dissolved in 100 ml of distilled water and left for 24 hours. 2 ml of dye extract was added and vigorously shaken in 2 ml of distilled water (DW), presence of persistent layer of bubbles indicated the presence of saponins. 5 ml was dilute ammonia solution was added to 2 ml of dye extract, two drops of conc. H_2SO_4 was added to this mixture, a yellowish colour revealed the presence of flavonoid. 2 ml of dye extract was added to 1 ml of Molisch's agent in addition to two drops of conc. H_2SO_4 and shaken together, a purple colour indicated the presence of carbohydrate (CHO). 1 ml of dye extract was added to 2 ml of 5% ferric chloride, a greenish colour revealed the presence of alkaloid [14].

2.3. *UV/VIS Spectroscopy*

1 g of *S.anisatum* dye extract was dissolved in 100 ml of methanol, 10 ml was measured into a Cuvet. The calibration and standardization of the equipment was done and methanol was used as the reference liquid before the UV/VIS of the extract was obtained [13-15].

2.4. *Scanning Electron Microscopy*

S.anisatum was prepared by spreading 20 g of the extract on a plate, this specimen was mounted on the stage set-up and observed when electron signals are sent and back scattered as electrons. The signals are scanned and received by the secondary electron detector, they are sent to a monitor and the information obtained is the micrograph representation of the sample [16].

2.5. *Photovoltaic Characterization of S.anisatum DSC*

An active area of 1.68 cm^2 of indium doped tin oxide (ITO) was exposed to insolation. TiO_2 of Degussa 99.8 % assay was uniformly blended with conc. HNO_3 acid, to which a few drops of cellulose acetyl was added to facilitate the surface cohesion. This smooth paste was applied by doctor blade method onto the ITO conducting side. This was sintered using Vecstar furnace at 450^0C for 1.5 hours. The result was a TiO_2 photoanode. The counter electrode was prepared by, coating the conducting side of a second ITO slide with soot over a naked Bunsen flame in a simulated vacuum enclosure. The slides were allowed to cool at room temperature, then 0.1 ml of dye extract is then deposited on the TiO_2 photoanode. The two conducting slides are fastened with crocodile clips, two drops of iodine solution and liquid n-type electrolytes are inserted in-between the sandwiched slides [17]. This set-up was connected in parallel with a resistive load and multimeter according to Ohm's law and exposed to irradiation under 1.5 air mass conditions.

3. Results and Discussion

3.1. *Phytochemical Analysis*

The phytochemical screening result revealed the following active chromophores: phenols, flavonoids, tannins, glycosides, alkaloids, saponins, terpenoids and protein. The pH of the solution was 5 and the most predominant phytochemical constituent is the phenolic content and its role is illustrated in Figure 1. It articulates with other functional ligands thus, extending the ancillary structure of *S.anisatum* in the DSC, this is the lengthening mode it uses to facilitate charge transport.

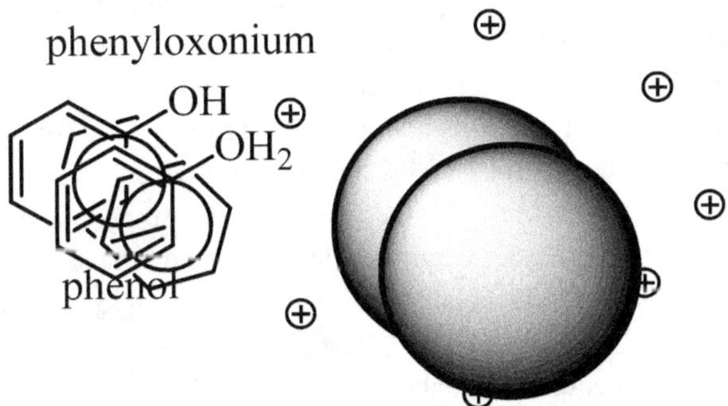

Fig. 1. Phenol chromophore's function in charge transport.

3.2. *UV/VIS Spectroscopy*

S.anisatum dye records peak absorbances at 350-400 nm and 680 nm respectively, this represents near ultraviolet and visible regions of the electromagnetic spectrum respectively as shown in Figure 2. Thus, the dye exhibits a porphyrin characteristic of absorption which implies that, it is absorbs optimally a large amount of solar energy. Previous studies show that the highest efficiency record was observed in porphyrin dyes.

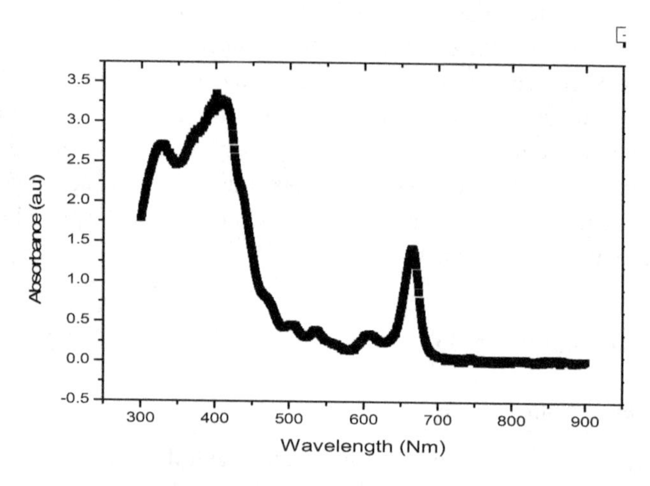

Fig. 2. UV/VIS of *S.anisatum* dye extract.

3.3. *SEM Spectroscopy*

The Scanning electron micrograph of *S.anisatum* reveals a multivariate non-crystalline structure as shown in Fig. 3(a). Using Gwwydion program software, it predicts the electron dynamics at the microstructural level. The red dots signify the electron shells of *S.anisatum* dye. There are more electrons in the conduction band, the mid-region than at the peripheral sides. The external layer depicts the valence band of *S.anisatum* dye, where fewer electrons occupy. This is a departure from the norm, usually the converse occurs in most substances. The implication of this is that, the phenol's aromatic property aids the electron transitions from the valence band to the conduction band. The blue area indicates where electron tunneling occurs, these are possible recombination trap sites. The upper red dots signify the excited shells where are fewer still due to the band gap energy required to make the transition [18].

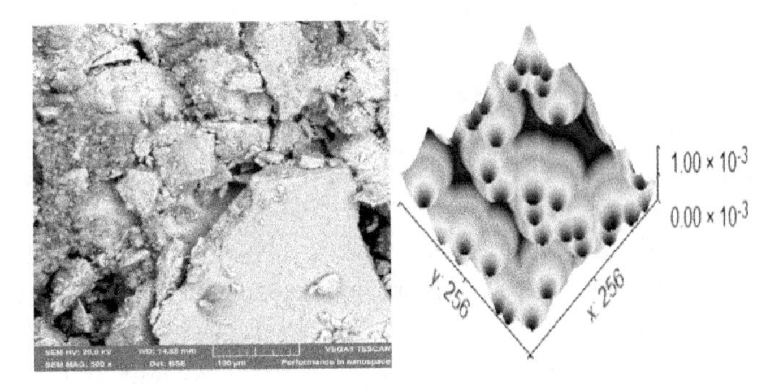

Fig. 3. SEM Micrograph of *S.anisatum* and the Gwwydion Plot.

3.4. *Photovoltaic Characterization*

The output performance of *S.anisatum* DSC was determined from the following parameters; short circuit current (I_{sc}) of 0.07 mA, open circuit voltage of 68.8 mV, fill factor (*ff*) value of 0.84 and the output efficiency (η) was 0.027 % as shown in Fig. 4. The *ff* and η were calculated from Equations 1 and 2 respectively. Where A is active area (in metres) of ITO exposed to insolation and P_{in} represents the power intensity of the radiation (in Wm^{-2}).

$$ff = \frac{P_{max}}{I_{SC}V_{OC}} \quad (1)$$

$$\eta = \frac{I_{SC}V_{OC}}{P_{in} \times A} \quad (2)$$

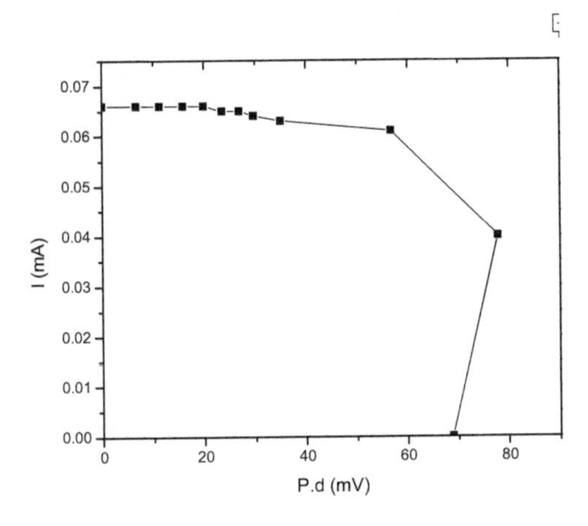

Fig. 4. I-V Curve of *S.anisatum* DSC.

3.5. *Bandgap Analysis on Schckley and Quiesser Limit*

The spectrum loses (U) that occurred in *S.Anisatum* DSC is interpreted from the Bandgap perspective of *S.anisatum* dye. A bandgap of 2.62 eV as shown in Fig. 5 implies that only phonons with higher HOMO-LUMO were absorbed from the electromagnetic environment surrounding *S.Anisatum* DSC. In relation to Fig. 3, the Shockley and Queisser factor [19-21] was determined by the thermal relaxation that resulted in the greater number of electron-hole pairs present at the peripheral and mid-section (81%) of the Gwwydion relative to the upper cut (18%). This is indicative of radiactive recombination from electrons in TiO_2 (bandgap of 3.2 eV) frame upon which *S.Anisatum* is based combining with the holes from the electrolyte. These mitigating loses eventually resulted in a huge power loss shown by the relative percentage of migration that occurred from higher occupied molecular orbital (HOMO) to lower unoccupied molecular orbital (LUMO). This is a sequel of the thermal relaxation in Fig. 3, thus the ultimate efficiency factor associated with spectrum losses, U is 0.22% in *S.Anisatum* DSC. This was obtained from a probability distribution function of the electron-hole pairs. The Fermi level of *S.anisatum* dye is therefore somewhere between 2.62 eV given by electron occupancy of the shells. The illustration of *S.anisatum* in Fig. 3 is not a simple bandgap, there is a wide difference between the electron density for the valence and conduction bands. This accounts for the wide margin between the Fermi potential (2.62 eV) and the open circuit voltage (0.07 eV). Thus, the redox potential of KBr electrolyte was not sufficient to overcome the energy difference of the quasi-Fermi level of the TiO_2.

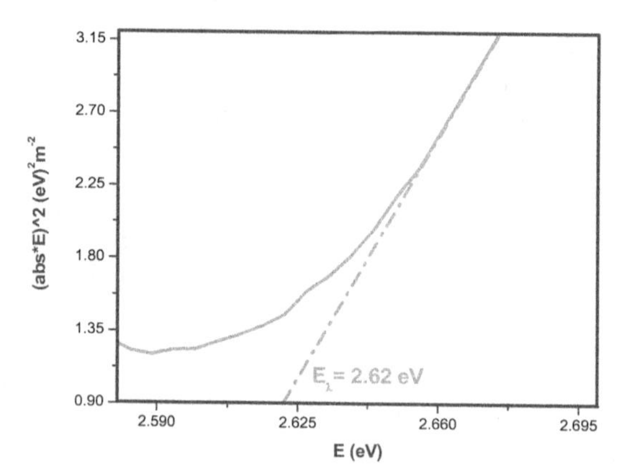

Fig. 5. Bandgap of *S.anisatum* Dye.

3.6. *Multiple Exciton Generation (MEG) in S.anisatum DSC*

S.anisatum dye will also be considered from the MEG perspective [21], *S.anisatum* organic dye has numerous electron states which at every time *t*, second is either occupied by an electron or is in an empty state. Some states require enormous energy to get filled up thus, at low temperature, they are usually scanty or empty. Other states which require low energy readily get filled up. In conventional semiconductors, there are a large number of states in both the valence and conduction bands. However, there are no states in the band gap. An organic semiconductor on the other hand, is characterized with an almost full valence band in which, the few empty states are holes. Its near empty conduction band, is occupied by a few conduction electrons. A perfect condition occurs when 100% of incident photons above the bandgap and zero beneath, are absorbed under ideal conditions of air mass. This incident photon causes photogeneration of electrons, causing a disruption of the erstwhile equilibrium balance in number of electrons and holes as illustrated in Fig. 6. A situation in which there is an increase in the number of electrons and holes in the conduction band. In order to offset non-equilibrium in the dye solar system, a photon exits through radiative combination. This is followed closely by net non-radiative generation resulting from perturbations and non-radiative recombination as energy gets converted into vibrations and heat in the dye- sensitized solar cell.

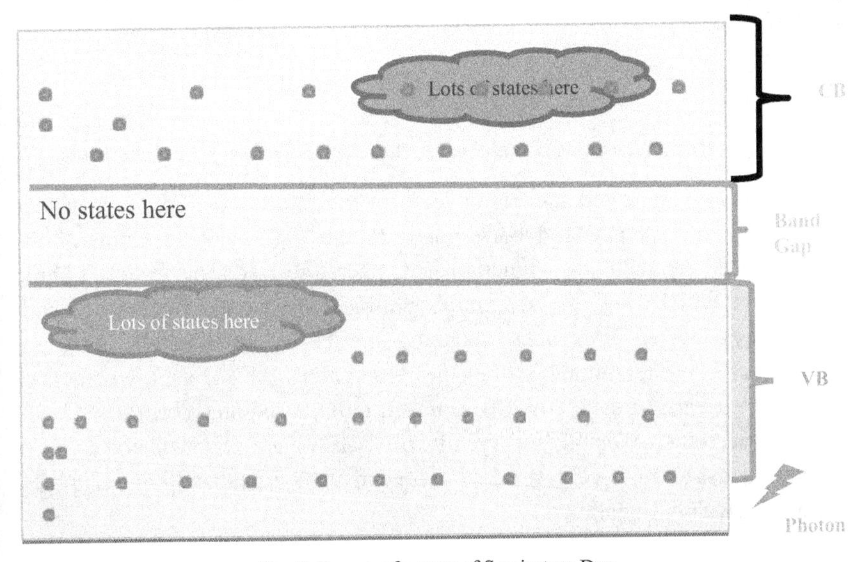

Fig. 6. Energy of a state of S.anisatum Dye.

The system is so desperate for equilibrium that, current will flow despite a negative voltage. Since the only way to eliminate radiative recombination is to fabricate a dye-sensitized solar cell that would not absorb light, it becomes imperative to discuss the net direct current generated by the dye cell which is given by Equation 3.

$$P\,(IV) = Photogeneration - Radiative\ recombination\ (V) \tag{3}$$

At higher voltages, more radiative recombination (*RR*) occurs in the dye-sensitized solar cell and the best case of *RR* is given by Equation 4.

$$RR = e^{\frac{qV}{K_B T}\frac{2\pi}{C^2 h^3}\int_{E_{gap}}^{\infty}\frac{E^2 dE}{\exp(E/K_B T_{cell})}} - 1 \tag{4}$$

Where *RR* is the radiative recombination (volts), q is the charge, V is the potential difference (volts), K_B is the Boltzmann's constant, T is absolute temperature (K), c is velocity of electromagnetic radiation in a vacuum (m/s), E is energy of the state.

The above discussion lends credence to the principle that efficiency of a dye-sensitized solar is directly proportional to the cost. Thus the total power P_{in} is a function of incident photons, the efficiency is higher when P_{out} is high and P_{in} is low. P_{in} is largely attributable to atmospheric conditions of latitude and longitude, angle of inclination of dye-sensitized solar cell all of which are relatively constant.

$$\eta = \frac{P_{out}}{P_{in}} \tag{5}$$

3.7. *Fermi-Dirac Perspective on S.anisatum Dye Efficiency Performance*

The multimeter indicated the Fermi level of the *S.anisatum* DSC which was equivalent to 68.8 mV and corresponds to the $V_{oc.}$ A consideration of the occupancy of the states as a function of temperature leads to the Fermi-Dirac distribution which presents several possibilities; more electrons at a higher Fermi level, presence of holes reducing the Fermi level of the state, an equilibrium where total number of electrons and holes is a constant, where there would be no Fermi Level. Another practical outlook requires consonance with MEG, a prerequisite for this theory. In MEG, each band is considered to be at different quasi-Fermi level (QFL). Consequently, the relationship of occupancy to state is given in Equation 6.

$$QFL = (electrons) \times (holes) \propto exp\frac{\Delta E}{K_B T} \tag{6}$$

This principle of QFL is based upon transition of electron-hole pairs, which is based on a consideration of Kinetic theory. It is subject to the degree of frequency at which an electron collides with a hole, which is radiative recombination as expressed by Equation 7.

$$RR \propto (electrons) \times (holes) \tag{7}$$

Therefore, Equation 6 becomes:

$$(RR) \propto \exp \frac{(QFL\ split)}{K_B T} \tag{8}$$

Thus, in the absence of resistance and if there is resistance, the condition is Equation 9 and 10 respectively.

$$QFL\ splitting = external\ voltage \tag{9}$$

$$QFL\ splitting\ > external\ voltage \tag{10}$$

The power breakdown in *S.anisatum* DSC is given by considering the value of the bandgap to the V_{oc} as illustrated by Equation 11.

$$Bandgap < External\ voltage \tag{11}$$

A photon with energy equivalent of the bandgap creates an electron-hole pair. A photon with higher energy than the bandgap would also create only an electron-hole pair. Therefore, the remaining energy is just waste. This implies that, there was a considerable massive energy loss in *S.anisatum* dye-sensitized solar cell.

4. Conclusion and Recommendation

1. *S.anisatum* DSC generated electricity from photons from the ultraviolet and visible regions of the electromagnetic spectrum.

2. *S.anisatum* absorbed radiant energy optimally but experienced power loses from thermal loses and radiactive recombination.

3. Future investigation into Electron bridging length and anchoring groups in *S.anisatum* is recommended as probe to boost output DSC performance [15].

4. The use of other electrolytes such as Acetronitrile with higher redox potential is highly recommended to improve on the open circuit voltage value.

The significance of this result is that, *S.anisatum* dye co-sensitization with other dyes with lower bandgap is favourable for higher efficiency. *S.anisatum* dye grafted on optoelectrical devices and fibers would also find applications in higher efficiency and in power saving devices.

Acknowledgments

The authors wish to express their gratitude to the technologists at department of Mechanical Engineering, University of Johannesburg, South Africa for their assistance with characterization of the samples. They also wish to convey their appreciation to the technologists in Physics laboratory of Covenant University for their assistance during the research. They are also grateful to Covenant University for the provision of the apparatus used for the completion of this work.

References

1. M. K. Nazeeruddin, E. Baranoff and M. Grätzel, Dye-sensitized solar cells: A brief overview. *Solar Energy* **85**, 1172–1178 (2011).
2. A. Malik and E. Grohmann, Environmental Protection Strategies for Sustainable Development, Strategies for Sustainability, *Dyes-Environmental Impact and Remediation* 114-117 (2012).
3. S. Ito, T. N. Murakami, P. Comte, P. Liska, C. Grätzel, M. K. Nazeeruddin and M. Grätzel, Fabrication of thin film dye sensitized solar cells with solar to electric power conversion efficiency over 10%, *Thin Solid Films* **516**, 4613–4619 (2008).
4. K. Kalyanasundaram and M. Grätzel, Artificial photosynthesis: biomimetic approaches to solar energy conversion and storage, *Current Topics in Biotechnology* **3**, 298-310 (2010).
5. P. Melhem, P. Simon, J. Wang, P. Simon, J. Wang, C. Di- Bin, B. Ratier, Y. Leconte, N. Herlin-Boime, M. Makowska-Janusik, A. Kassiba and J. Boucle, Direct photocurrent generation from nitrogen doped TiO_2 electrodes in solid-state dye-sensitized solar cells: towards optically-active metal oxides for photovoltaic applications. *Solar Materials and Solar Cells*, **117**, 624-631 (2013).
6. D. B. Menzies, Q. Dai, L. Bourgeois, R. A. Caruso, Y. Cheng, G. P. Simon and L. Spiccia, Modification of mesoporous TiO_2 electrodes by surface treatment with titanium (IV), indium (III) and zirconium (IV) oxide precursors: preparation, characterization and photovoltaic performance in

dye-sensitized nanocrystalline solar cells, *IOP Nanotechnology* **18(12)**, (2007).

7. H. Masood, J. Vahid and G. Afsahar, Dependence of energy conversion efficiency of dye sensitized solar cells on annealing temperature of TiO_2 nanoparticles, *Material Science in Semiconductor Processing*, **15(4)**, 371-379 (2012).
8. S. Mathew, A. Yella, P. Gao, R. Humphry-Baker, F. E. CurchodBasile, N. Ashari-Astani, I. Tavernelli, U. Rothlisberger, M. K. Nazeeruddin, M. Grätzel, M, Dye-sensitized solar cells with 13% efficiency achieved through the molecular engineering of porphyrin sensitizers. *Nat Chem* **6 (3)**, 242-247 (2014).
9. M. Zhang, Y. Wang, M. Xu, W. Ma, R. Li, P. Wang, Design of High-Efficiency Organic Dyes for Titania Solar Cells based on the Chromophoric Core of Cyclopentadithiophene-Benzothiadiazole. *Energy & Environmental Science* **6 (10)**, 2944-2949 (2013).
10. Y. Jiao, F. Zhang, M. Grätzel, S. Meng, Structure–Property Relations in All-Organic Dye-Sensitized Solar Cells. *Advanced Functional Materials* **23 (4)**, 424-429 (2013).
11. J. H. Delcamp, A. Yella, T .W. Holcombe, M. K. Nazeeruddin, M. Grätzel, The Molecular Engineering of Organic Sensitizers for Solar-Cell Applications, *Angewandte Chemie International Edition* **52**, 376-380 (2013).
12. A. F. Bartelt, R. Schuetz, C. Strothkämper, J. Schaff, S. Janzen, P. Reisch, I. Kastl, M. Ziwritsch, R. Eichberger, G. Fuhrmann, D. Danner, L. -P. Scheller, G. Nelles, Efficient Electron Injection from Acyloin-Anchored Semi-Squarylium Dyes into Colloidal TiO_2 films for Organic Dye-Sensitized Solar Cells, *The Journal of Physical Chemistry C* (2014).
13. T. Abodunrin, A. Boyo, M. Usikalu, L. Obafemi, O. Oladapo, L. Kotsedi, Z. Yenus, M. Malik, Microstructure characterization of onion (*A.Cepa*) peels and thin films for DSSCs. *Materials Research Express* **4**, (2017).
14. T. J. Abodunrin, A. O. Boyo, M. R. Usikalu, Data on the porphyrin effect and influence of dopant ions on Thaumatococcus daniellii dye as sensitizer in dye-sensitized solar cells, *Data in brief* **20**, 2020-2026 (2018).
15. T. J. Abodunrin, O. O. Ajayi, M. E. Emetere, A. P. I. Popoola, U. O. Uyor, and O. Popoola, "Electron Tunneling, Performance Analysis and Prospect of Micro-energy Generation in Ringwood (Syzygium Anisatum) Dye-sensitized Solar Cell," Lecture Notes in Engineering and Computer Science:

Proceedings of The International MultiConference of Engineers and Computer Scientists 2019, 13-15 March, 2019, Hong Kong, pp322-325.

16. C. Longo and M. A. de Paoli, Dye-sensitized solar cells: a successful combination of materials, *Journal of the Brazilian Chemical society* **14(6)**, 889-901 (2003).
17. B. E. Hardin, H. J. Snaith and M. D. McGehee, The renaissance of dye-sensitized solar cells. *Nature Photonics* **6**, 162-169 (2012).
18. L. U. Okoli, J. O. Ozuomba and A. J. Ekpunobi, Influence of local dye on the optical band-gap of titanium dioxide and its performance as a DSSC material. *Research Journal of Physical Sciences* **1**, 6-10 (2013).
19. H. Nguyen, U. Gottlieb, A. Valla, D. Munoz, D. Bellet and D. Munoz-Rojas, Electron tunneling through grain boundaries in transparent conductive oxides and implications for electrical conductivity: the case of ZnO: Al thin films, *Materials Horizon* **4**, (2018).
20. M. Zhang, Y. Wang, M. Xu, W. Ma, R. Li, P. Wang, Design of High-Efficiency Organic Dyes for Titania Solar Cells based on the Chromophoric Core of Cyclopentadithiophene-Benzothiadiazole, *Energy & Environmental Science* **6 (10)**, 2944-2949 (2013).
21. W. Shockley, H. J. Queisser, Detailed Balance Limit of Efficiency of P-N Junction Solar Cells, *Journal of Applied Physics* **32**, 510-519 (1961).

CPSIA information can be obtained
at www.ICGtesting.com
Printed in the USA
JSHW011433090120
3448JS00001B/8